P9-EMP-025

Fire, Ice, and Physics

Fire, Ice, and Physics

The Science of *Game of Thrones*

Rebecca C. Thompson

Foreword by Sean Carroll

The MIT Press
Cambridge, Massachusetts
London, England

MORRILL MEMORIAL LIBRARY
NORWOOD, MASS 02062

500
Thompson

24.95

© 2019 Rebecca C. Thompson

All rights reserved. No part of this book may be reproduced in any form by any electronic or mechanical means (including photocopying, recording, or information storage and retrieval) without permission in writing from the publisher.

This book was set in Stone Serif and Trajan Pro by Jen Jackowitz. Printed and bound in the United States of America.

Library of Congress Cataloging-in-Publication Data

Names: Thompson, Rebecca C., author.
Title: Fire, ice, and physics : the science of Game of Thrones / Rebecca C. Thompson ; foreword by Sean Carroll.
Description: Cambridge, MA : The MIT Press, [2019] | Includes bibliographical references and index.
Identifiers: LCCN 2019001209 | ISBN 9780262043076 (hardcover : alk. paper)
Subjects: LCSH: Science--Popular works. | Science in popular culture. | Game of Thrones (Television program)
Classification: LCC Q162 .T42945 2019 | DDC 500--dc23
LC record available at https://lccn.loc.gov/2019001209

10 9 8 7 6 5 4 3 2 1

To

P.W.H., E.J.E., and R.I.M.

Next time I have an idea like that, punch me in the face.

—Tyrion Lannister

Contents

Acknowledgments

First and foremost, I'd like to thank my parents, Faye H. Daniel and Dr. Steven D. Thompson. Neither of you are here to see this completed, but your love and support throughout my life got me here, and I know you were looking down on me through the process. I absolutely would not have gotten through this on my own. Thank you, Amy Stern, for helping me figure out how this all works. Molly Kleinman, thanks for being an amazing research assistant and finding even the gory papers. Who knew you'd get to learn about hanging! Carolyn Kuranz, thank you so much for all your help with physics and graph production. Ted Beyer, my military historian friend, thank you for always having the answers and always making me think hard about mine. Thanks, James Roche and Stephen Skolnik, for making everything run smoothly and helping me keep a positive attitude. Thanks, too, for listening to an endless number of "BecX Talks" about everything from incest to Damascus steel. Many thanks to my copy editor, Elizabeth Agresta, for doing an amazing job and adding some fabulous jokes. A big thanks to *NCIS* for being my escapist show while writing about my escapist show. It has a slightly lower body count and at least Ziva and Tony weren't related—though I think I'd rather face Jon with a sword than Ziva with a paper clip. And of course, thanks to Bo, who has put up with more than any partner should.

FOREWORD

SEAN CARROLL

As I'm sitting down to write this foreword—somewhat after the book itself has been finished—a study was just published in the journal *Injury Epidemiology*, with the title "'Death Is Certain, the Time Is Not': Mortality and Survival in *Game of Thrones*." The authors, Reidar Lystad and Benjamin Brown, address the pressing question of what kind of survival strategies were most effective among the major players in Westeros. (It's a violent world; 14% of characters on screen died within an hour of their first appearance.) Their paper includes paragraphs such as this:

> Important characters appearing in Seasons 1 to 7 of *Game of Thrones* were included, and data on sociodemographic factors, time to death, and circumstances of death were recorded. Kaplan-Meier survival analysis with Cox proportional hazard regression modelling were used to quantify survival times and probabilities and to identify independent predictors of mortality, respectively.

Serious stuff! Or is it?

Scientists pride themselves on studying the real world. The world of George R. R. Martin's *A Song of Ice and Fire* books, and the TV show on which they are based, is not the real one. Martin made it up. Admittedly, certain situations and events were inspired by real-world history, but the *Game of Thrones* milieu features entirely fabricated climatology, astronomy, metallurgy, chemistry, and biology, not to mention zombies and dragons. What can science say about that?

Quite a bit, as you are about to discover.

The conversation between science and literature (science fiction, fantasy, or any other genre, for that matter) is a dialogue. Literature can learn from science in obvious ways. If your story is set in outer space, you're going to want to know how rockets and closed ecologies work. Even if your story is set in a feudal society suffused with magic, all sorts of science might be relevant, from weather patterns to the chemistry of various poisons.

But information and inspiration also flow in the other direction. Scientists collect data by doing experiments and observations, and use that information to gain knowledge of how the world works. Reading a work of fiction is data collecting of a sort. If the world of the story is well constructed, it will obey rules, whether or not those rules are explicitly laid out. If anything goes, the story isn't interesting; for the protagonists to be challenged and the audience to be engaged, characters have to operating within a logical milieu. Without physics, there can be no drama. A good scientist can examine a well-told story and figure out what the rules of the world are, whether or not they're the same of our world. That's what scientists do.

Here in *Fire, Ice, and Physics*, you'll encounter a masterful exploration of both sides of the dialogue between science and fantasy. Rebecca Thompson takes the world of *Game of Thrones* and examines it through the eyes of a trained scientist. If there's one thing that everyone who watches *Game of Thrones* knows, it's that "winter is coming"—but on a somewhat unpredictable schedule. Unlike Earth, where we can predict well in advance when the leaves will start turning and the temperature start dropping, seasons in Westeros are much sneakier. It would be okay to think to yourself, "Well, GRRM just made that up for dramatic effect; there's nothing scientific about it." And you might be right. But when faced with that kind of unusual phenomenon, a scientist can't help but think, "But how *would* that work . . . ?"

Examining questions like that—and not necessarily *answering* them—is what makes this book so delightful. Answers can be hard to come by, and since we can't visit Westeros to collect additional data, we may never know for sure. But science is a process, not just a set of established results. And here you will see that process at work. Once we apply our brains to the problem, rather than just dismissing it as a fictional conceit, we quickly realize the incredibly rich set of scientific concepts that can usefully be brought to bear.

Happily, Martin's world has given us an enormous amount of raw material to work with. The books, as well as the TV series based on them, are famously detailed, from what the smallfolk tend to eat for typical meals to what the mottos are for a dizzying number of noble houses. The scientific questions are similarly numerous and rewarding.

Of course, most of the questions seem just as fantastical and hopelessly unscientific at first glance. The Wall in the North is held up by magic, we are told explicitly. Nobody even tries to explain Valyrian steel, or how dragons can breathe fire. And wildfire is simply presented as an incredibly dangerous substance, not the careful product of diligent lab work by Westerosi chemists.

But here's the thing about science: it's always there, lurking beneath the surface. *Game of Thrones* is fantasy, but it's not surrealism; everything that happens is either based on or inspired by features of our actual world. Take the example of the Strangler, the poison used to kill King Joffrey at the Purple Wedding. There's no reason why it has to be an actual poison we have here on Earth, but it has properties we know and can analyze: it has to be something we can disguise as a gem on a necklace; it dissolves in wine; it doesn't have a strong taste; it constricts the throat and renders the victim incapable of breathing. Scientifically speaking, we're given a lot of data to work with.

As Thompson shows, there's no known poison that matches precisely what we're told about the Strangler, but we can come pretty close. Strychnine, in particular, can be made into the form of a crystal, and it kills by causing muscle contractions. But it's not a perfect fit, since it affects muscles all over the body, not just in the throat.

That's okay. The point is not to find a perfect fit; *Game of Thrones* is fantasy, not a documentary. What matters is that by investigating the question, we end up learning something along the way. If all you did was watch the TV show, you'd be left with a feeling of satisfaction at seeing a vicious, immature monarch brought down. But you wouldn't necessarily appreciate that a strychnine-like poison does its damage by blocking the neurotransmitter glycine, therefore causing electrical signals in the brain to go haywire and resulting in a rapid death. That's probably something that not even the Maesters of the Citadel really understand, but this book is here to fill you in.

What makes a book like this so much fun is just how many examples there are of fantastical events or objects that can teach us something about science. Even if the Wall is held up by magic, what do we know about the structural properties of ice that can tell us exactly what kind of magical assistance would be required? Dragons don't exist, but dinosaurs did, and their biology and evolution can tell us something about what dragons might be like. (Interestingly, Martin's dragons have two legs and two wings, which differs from the traditional four-legged dragon of mythology, but is biologically more realistic.) White Walkers are make-believe, but the natural world has some species that have more in common with zombies than you might think.

Like all good fiction, what happens in *Game of Thrones* is driven by the goal of telling a good story, not by being scientifically accurate. But the spirit of science is useful in any situation, and that spirit comes vividly to life in *Fire, Ice, and Physics*. As much fun as it is to kick back and lose ourselves in an alternative fictional reality, there's an extra dimension of enjoyment we get from thinking about what we're watching in a scientific way.

And who knows? We might learn something useful. After all, winter is coming.

Introduction

Because I'm a scientist, that's why. As a trained scientist, I ask lots and lots of questions, and I want to know why things happen the way they do. I want explanations and reasons for what I observe. I wish I could turn this need on and off, but unfortunately, I can't. What this means is that watching TV and reading fiction can be an interesting endeavor that often leads to me hit pause and yell to my fiancé about the questionable science presented. We then have the joy of a 20-minute back-and-forth on science, as we used to do in grad school. I hope that my obsession with asking so many questions and needing the answers will not ruin your fantasy fun but rather enhance it. I'm not out to nitpick every small detail in every scene in *Game of Thrones*. I don't want to look at every sword swing or arrow trajectory and tell you why that couldn't happen. So much of what's done in Hollywood is done for drama, and I don't want to spend too many pages telling you why it's wrong. What I want to do is use *Game of Thrones* as a gateway to learning some really interesting science, and then use this knowledge to add a new dimension of appreciation to a really great show.

My day job, when not writing about the science of Hollywood, is getting nonscientists to appreciate science. My standard joke is that when I meet a guy in a bar and he asks me what I do, my answer is highly dependent on how attractive he is. If I want to continue the conversation, I say I write science-based comic books. If I'd prefer he leave, I say I'm a physicist. Both are technically true, though one is much less intimidating. My goal in life is to change that interaction.

This book began as a talk I gave at Biosphere 2 in Arizona. No, not *Biodome* (a horrible movie starring Pauly Shore)—rather, the experiment in which people were sealed in a glass dome to see if they could survive for three years. (They could not.) It's now a research facility and an institute for training science teachers. I would highly recommend a visit, but make sure you go off the path and see the ruins. The organizers invited me to teach some classes and give a casual physics lecture during dinner. I asked what they wanted me to talk about, and they replied, "Anything you want." I had written blog posts on the physics of various things, but I had never been able to give a full talk on "anything," and I was not about to waste it. *Game of Thrones* had just become the number one pirated show in the world, and I was hooked. I saw so much interesting science in the show, and it seemed like a good "anything" to try. What I didn't expect was that I'd be walking into a room with a talk full of blood and gore and death and graphic video clips—and that I'd be the only person who had seen the show. In spite of this, two things happened. Lots of people learned a lot of science, and more than a few new fans were converted. I'm not entirely sure the person who invited me to give this talk fully understood what I would interpret as "anything," but here we are, many iterations of that talk—and now, a book—later.

Because this began as a talk, and because I am most comfortable explaining things as if I'm talking directly to a real person, my style is very conversational. In grad school, my advisor would push us to explain our research "as if you were explaining it to your mother." I pointed out this was more than a little sexist, and that my officemate's mother had a degree in engineering and could probably handle the big words. He rephrased, asking us instead to explain it to a 12-year-old. It forced us to find the core ideas of our work and explain those only. I can't thank him enough for making us do that. It made us better scientists, better teachers, and better at applying for grants. I don't know if my narrative choices will be good or bad, but I felt like it was how I do things, so why not?

I firmly believe everyone should have an understanding of at least a little science and be comfortable using common terminology and thinking the way a scientist would in order to understand the natural world (and occasionally the unnatural ones). So often you hear people say, "I'm not really a science person," or "I'm not really good at science," yet it's not socially acceptable to not really be "a reading person" or "good at politics." These are all skills that people need to learn to navigate society, and

I don't think science should be any different. There are different methods to get people interested in and comfortable with science, and through my job I've had the opportunity to try several, ranging from blog posts to comic books. I am hoping that this foray into a piece of Hollywood science will reach people in a way I haven't yet tried. I hope you enjoy it, but more than anything, my goal is for you to look at things in a different light by the end of the book. There will be spoilers. (So, so many spoilers.) If you haven't watched through season 7, know that you will learn what happens to key characters. You have been warned.

What I don't want to do is take away your enjoyment of the show. Season 7 was airing when I began this project, and I was so worried I would only be able to watch the show as a scientist and that I would lose what I so loved about it: the characters and the interplay between them, their relationships, the unexpected deaths, and the dragons. I realized my fears were unfounded when I teared up as Daenerys rode in for the first time on Drogon with sweeping wings and flame-torch breath. I sobbed when Viserion was killed and cheered when Jon and Dany *finally* got together. My emotional involvement with the show actually increased rather than decreased, and I hope the same will be true for you. I hope this book will give you a deeper understanding of how rich and deep the world of Westeros really is.

I had a lot of fun picking topics I wanted to focus on and identifying specific pieces of the show that had fun scientific explanations. Because so much of the show involves death, I knew that would be a key chapter. I figured I'd watched so much death on the show that it couldn't be too hard to write about the science of dying. I was very, very wrong. It was emotionally difficult, and despite my best efforts, I was not able to separate myself as a scientist as much as I'd have liked to. Know that it might be harder to read than you may have expected. But in writing that chapter, I learned two things. The first is that being a blonde woman at a cocktail party coherently discussing the science of the guillotine gives some people a shock and causes them to dash off for another drink. The second is that there is no easy way to go from alive to dead. It's a hard transition no matter what, but I guess we'll all find that out eventually. I chose not to go into the science of various methods of torture because that seemed like a bridge (or rat in a bucket) too far.

I know there are more than a few internet forums and pop science articles devoted to many aspects of both the show and the world within. Dragon fire and wildfire stand out as particular topics people like to discuss. In as many cases as possible, I have tried to address these elements of the show or use them as a starting point for discussions about real-world science. If you find yourself getting agitated and disagreeing with me, please give me a chance and read the full argument. I have used primary source references and a lot of scientific reasoning to back up my work. In some cases, I may very well have missed an argument or a potential scientific explanation, but I am certainly excited to learn about them. Publisher willing, perhaps there will be a second edition that can address these issues and potentially answer all of the interesting science questions you still have. HBO is already casting for a new prequel series, so who knows what will happen!

I hope I was able to make the science of these really complicated topics easy to understand. What I talk about here is what I thought was the most relevant to the topic at hand, not the sum total of the information out there. I've had the interesting experience of finding myself talking intelligently about the myth of saltpeter being used to suppress sex drives of soldiers or the efficiency of the German guillotine, but I didn't include either in the book since they have little relevance to Westeros. Thanks to modern social media, however, I am *more* than happy to have lengthy discussions about all of these things.

More than anything, read and have fun. Learn something and become your own cocktail party buzzkill or the center of attention, depending on the crowd. Enjoy the show. Enjoy the science. Have fun, get emotional, get involved, and let science add to your fun. I hope you like this book as much as I liked writing it. I learned a lot and I hope you do, too.

All this being said, let me paraphrase Sansa Stark and Jon Snow: Science is coming. I have been promising.

1

Winter Is Coming—Or Is It?

Seasons in Westeros

"Oh, my sweet summer child," Old Nan said quietly, "what do you know of fear? Fear is for the winter, my little lord, when the snows fall a hundred feet deep and the ice wind comes howling out of the North."
—A Game of Thrones

"Winter Is Coming"—these are both the words of House Stark and George R. R. Martin's ominous refrain throughout *A Song of Ice and Fire*. Behind the fighting of kings and the birth of dragons is that promise: winter is coming. Well, of course it is. It's usually the season that follows fall and precedes spring. Why are the Starks making such a big deal of this? Because this isn't taking place on Earth, that's why. The continent of Westeros is on a planet that, for the most part, operates the same way as Earth. There's land and sea, the surface temperature seems about the same as Earth, and there are, well, humans. But as much as Westeros seems like Earth, it's clear from the early on in the series that something is amiss, and that something bigger than the fighting of kings is driving the plot. After all, political leaders fighting while dragons breathe fire isn't anything new. Doing it with the imminent threat of a years-long winter so cold it can kill while also unleashing an army of the undead makes the whole story that much more ominously interesting.

Before I go into what causes seasons and whether or not a planet like the one in *Game of Thrones* could exist, it's important to note that the biggest factors in Earth's average temperature are its atmosphere and the amount of the planet that is covered by water. It takes a lot more energy to heat and cool water than it does to heat and cool land. Because 71% of

Earth is covered by water, the average temperature of the planet doesn't swing that wildly from night to day and from season to season. Water simply cannot change temperature that quickly. Our atmosphere also protects us from temperature swings. Most of the energy from the sun that reaches Earth is trapped as heat in our atmosphere due to the greenhouse effect. Yes, the greenhouse effect is responsible for global warming, but it is also the reason our planet can sustain life. I'm going to assume that the planet in *Game of Thrones* has the same general properties as Earth, a similar atmosphere, and a similar percentage of ocean. It's possible that this is a bad assumption to make and that the unpredictable seasons are caused by swings in greenhouse gases or a lower percentage of water, but it seems safe to assume, based on the mythology, that this is not the case. That means there has to be some astronomical explanation—that is, an explanation having to do with the relationship between the planet and other massive bodies in its vicinity, such as moons and stars, for the seasons that drive the narrative of Westeros's foreboding onslaught of winter.

What Exactly Are Seasons?

It seems silly to answer this question at the start, since everyone pretty much knows what a season is. If flowers are blooming, it's spring; if it's hot and steamy, it's probably summer; and if the leaves are changing color, it's probably fall. Days off work and a snow shovel mean winter is finally here. In the world of Westeros, it's the white raven of the Citadel, not the French Toast Index,[1] that tells us when winter is here; however, it isn't quite that simple. Even scientists don't have one specific, failsafe way to say what season we are in. There are actually two working definitions of seasons, meteorological and astronomical. As you would expect, the meteorological definition is based on what the weather is like, and the astronomical definition is based on where we are in our orbit around the sun. In the Known World of Westeros, the maesters seem concerned primarily with the meteorological definition, but an interesting book in the restricted section of the Citadel's library seen in the season 7 premiere suggests they knew a bit more about astronomy than previously discussed. The two don't differ that much in terms of start and end dates, but for most of this chapter I'm going to be using the astronomical definition because I care more about how the planet is moving and less about rain next Tuesday.

Meteorologists divide the calendar into four sections called seasons. Given that I'm writing this book in the Northern Hemisphere, the show is filmed in the Northern Hemisphere, and GRRM lives in the Northern Hemisphere, I'm going to talk about seasons from that perspective. In the Southern Hemisphere, the seasons are reversed. In the North, meteorological winter is December, January, and February; spring is March, April, and May; summer is June, July, and August; and fall is September, October, and November. This isn't really surprising and generally aligns with our view of the seasons. These divisions were decided upon to make weather forecasting easier and to simplify the calculation of climate averages and trends, and they're definitely biased toward the mid-latitudes, where the seasonal differences are most noticeable. When you hear phrases like "hottest summer on record," it is referring to this definition of seasons. Figure 1.1 shows the average temperature by month for Washington, DC. If you look at the graph you can see that spring and fall really do represent the transition points. The average temperature is changing much faster during these six months than during summer and winter. The more

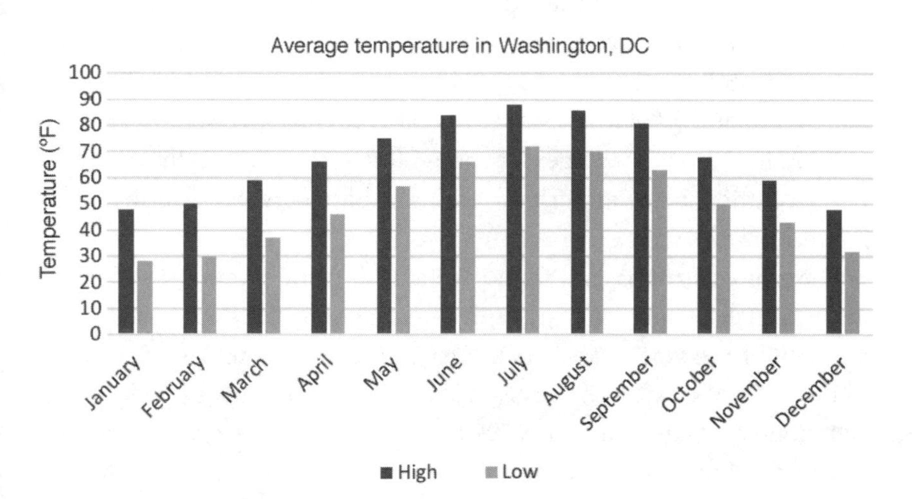

FIGURE 1.1
Average monthly high and low temperatures in Washington, DC. Data provided by the National Renewable Energy Laboratory.

math-y way to say this is that the derivative of the curve in spring and fall is greater than that in summer and winter. The derivative is a way to show how fast something is changing, so it makes sense to define seasons like this; however, this definition relates more to how humans have constructed months than how Earth is rotating around its axis and orbiting the sun. That's where astronomical seasons enter the picture.

Astronomical seasons lag approximately 20 days behind meteorological seasons. Instead of being based on what the weather feels like here on Earth, it's based on where Earth is in its orbit around the sun. There are four important points that mark the change from one season to another: the winter solstice, the vernal equinox, the summer solstice, and the autumnal equinox. You can probably guess by their names which point is associated with each season, but just in case, here's the official astronomical definition of seasons: Winter begins at the winter solstice, which is around December 21, and ends at the vernal equinox, around March 20, when spring begins. Spring lasts until the summer solstice, usually around June 21. Summer ends with the autumnal equinox, on or around September 22, and the cycle repeats itself. This definition of the seasons is based on two astronomical phenomena—the equinox and the solstice—so what exactly are those?

Both the solstice and the equinox were observed and named before Nicolaus Copernicus saw the publication of his heliocentric model of the universe on his deathbed in 1543. This means that the equinox and solstice were named for what people on Earth saw happening to the sun's position in the sky. In the next section, I'll talk about what this means in a heliocentric model, but for now, this is all based on observations of the sun from Earth.

The word solstice comes from the Latin *sol*, which means "sun," and *sistere*, which means "to stand still." As the sun moves across the sky throughout the year, it appears higher in the sky as the days get longer and lower in the sky as the days get shorter. If something is going from a high point to a low point and back again, it needs to turn around at two different points. These two points are called the solstices. The summer solstice is the longest day of the year, when the sun is at its highest point in the sky. From the winter solstice onward, the days become progressively longer and the sun rises progressively higher in the sky at midday.

On the summer solstice, it appears—or at least it did to ancient Greeks, anyway—that the sun stands still in its progression upward before turning around and heading back down to its low point on the winter solstice.

The equinoxes, meanwhile, fall at the halfway points between solstices. At these points, the number of daylight and nighttime hours are equal—hence the term *equinox*. You may have heard of the Tropic of Cancer and the Tropic of Capricorn, two important latitudes with respect to the equinoxes. (They are also fabulous books by Henry Miller.) On the vernal equinox, the sun is directly overhead at noon if you are standing at the Tropic of Cancer. On the autumnal equinox, this noonday sun occurs at the Tropic of Capricorn.[2]

Interestingly, many people think the amount of daylight time increases and decreases in a linear fashion; that is, it changes by the same number of minutes each day. This isn't the case. The further you are from the solstice, the faster the rate of daylight change. It forms a sine wave. Obviously, the meteorological seasons depend on the astronomical seasons; it's because of the change in how the sunlight hits Earth that the temperature on Earth changes. Though meteorological seasons are ultimately what the Westerosi care about, the rest of this chapter will talk about how the planet and its orbit may be causing the wonky meteorological seasons the inhabitants of Westeros and Essos are experiencing.

WHY DOES EARTH HAVE SEASONS?

Now that we've defined what seasons really are, we can look at what's causing them. If you are ever looking for something funny other than cat videos on the internet, search for "Harvard grads explaining why we have seasons." Their explanations indicate a few things: first, it is clear that physics is not a Harvard graduation requirement; and second, there are a lot of misconceptions about what causes seasons. Hopefully after reading this chapter you can answer the question better than they do.

When asked what causes seasons, most people will say that Earth is closer to the sun in the summer and farther from the sun in the winter. This isn't a bad assumption. The closer you are to a heat source, the hotter it feels, so it makes sense that being closer to the sun would make Earth's surface hotter. However, this assumption is wrong. Earth is actually closer to the sun in the winter and farther from the sun in the summer. It's true that Earth is not the same distance from the sun at all points in its orbit, but the difference in distance is very, very small. The change in distance does have an effect on the temperature—just not enough to cause seasons.

Most planets have an elliptical rather than circular orbit. Elliptical orbits have two extremes: the point when the planet is closest to the sun and the point when it farthest away. These are called the *perihelion,* meaning "near the sun," and the *aphelion,* meaning "away from the sun." For some planets, the difference between the perihelion and the aphelion is very large, but for the Earth, not so much. The difference between the two is only about 3.3%. When Earth is at its closest, it is 147,098,074 km away from the sun; at its farthest, it is 152,097,701 km from the sun. Yes, that's a difference of roughly 5 million km. This seems huge, but it's really only a small percentage of the total distance. This does, however, effect a small change in the temperature of the planet—a fluctuation of about 10%, in fact. When talking about thermal radiation from something like the sun, the temperature felt by something some distance away is proportional to the square of the distance. The temperature felt decreases fairly quickly the farther away you move from the heat source. This would suggest that the difference has a pretty big effect, but if you do the math, it really isn't that big—a 3.3% change in distance corresponds to about a 10% change in temperature. This means the difference in temperature due to the difference in distance is only about 10% of the average temperature of Earth, or about 5°F (2°C).[3] You now officially know more than most of Harvard's graduating class.

So, if seasons aren't caused by a change in distance from the sun, then what causes them? It turns out the tilt of our Earth's axis is responsible. This is an explanation you might have heard before, and if you type "why do we have seasons" into Google, this will be the answer you find. In most cases, however, the discussion is pretty cursory, and you are left to believe it without much math to back it up. I hope to do better than that. As I said before, I live in Washington, DC, so I'm going to use that as an example. We have four seasons that are pretty regular, so it seems like a good choice. The average temperature in the summer is about 80°F (27°C) and in the winter it's usually about 35°F (4°C).[4] Something about Earth's motion is causing a 45°F (23°C) swing in temperatures and it can't be the orbit around the sun. As the Earth moves around the sun it spins on its axis like a top. But unlike a top, it doesn't spin straight up and down, the axis around which it spins is about a bit of an angle—about 23.5°, to be more accurate. The tilt means that sunlight isn't hitting the Earth straight on at all times in its orbit. I remember being 10 years old and learning that Earth's axis was tilted, and the first image that came to mind was that of the axis always tipped slightly toward the center of its orbit—always tilted toward the sun, like if you swung a ball on a string. My 10-year-old self could not understand why that caused seasons. I also needed a better picture of what it meant to be tilted 23.5°. Earth doesn't orbit the sun like a top and the axis doesn't act like a string. The axis is always pointed in the same direction relative to the plane of its orbit, so sometimes it's tilted toward the sun, sometimes away (figure 1.2).

It's the tilt—more specifically, the fact that sometimes it's tilted toward the sun and sometimes it's tilted away from the sun—that causes seasons. When the Northern Hemisphere is tilted toward the sun, it receives direct sunlight; when it is tipped away, the sun's rays hit the surface at an angle.

FIGURE 1.2
Earth moving around the sun. The axis stays tilted in the same direction as it rotates around the sun. Sometimes Earth is tilted toward the sun and sometimes it's tilted away—this is what causes seasons.

That angle dictates the intensity of the sunlight, which in turn affects the temperature on Earth. When I first heard this, my 10-year-old self was not convinced. How could the angle of sunlight make that much of a difference? Hopefully some math will convince you.

Earth's temperature is determined by how much of the sun's radiation reaches the surface. The more radiation per square kilometer Earth receives, the warmer it will be. To me, it didn't seem like a slight tilt would do all that much; however, my intuition was completely wrong on this point. Think of an amount of sunlight hitting a square kilometer from straight overhead. That square kilometer receives a certain amount of energy per second from the sun, and the amount of energy received determines the temperature. Now imagine that the sunbeam is tilted slightly (i.e., lower in the sky). In fact, don't imagine—go get an old-school flashlight and see for yourself. You can do it with an LED flashlight, too, but the beam isn't as easy to see. A flashlight emits a fixed amount of energy per second, just like the sun. Take the flashlight and shine it on a surface head-on. Look at the size of the spot and see how bright it is. Now tilt the flashlight a bit. You'll see that the spot is a whole lot bigger, and that it's now elliptical (oval-shaped) rather than circular. You'll also notice that it doesn't look as bright. The same amount of energy is in that beam of light, but now it's hitting a larger area. So, because the same amount of energy is spread out over a wider area, each spot on the table or wall (or wherever you are shining your flashlight) is getting a whole lot less energy per second. If the flashlight is tilted at a 45° angle, the light spreads out over an area 1.4 times the size of the original spot. That's a 40% increase in area. At a 60° angle, the area is increased by 67% (1.67 times larger than the original). In the case of Earth, it's the table tilting instead of the flashlight, but the effect is the same—the intensity decreases as the tilt increases. And because the Earth isn't flat, the change in intensity (and therefore temperature) isn't spread out evenly. In addition, places farther from the equator experience more dramatic seasonal temperature changes than places closer to the equator do. The temperature in, say, Miami changes less throughout the year than it does in a place like DC because Miami is closer to the equator.

Think of the sun shining over Washington, DC. In the summer, that sunlight hits much more directly—and therefore more intensely—than it does in the winter because Earth's Northern Hemisphere is tilted toward rather than away from the sun. The more direct the sunlight, the higher the average temperature. The National Renewable Energy Laboratory measures exactly how much sunlight hits a certain spot on Earth at a certain time and calculates the corresponding temperature. You can find this information in the National Solar Radiation Database (NSRD). Every 30 minutes, the database measures the amount of sunlight on one square kilometer of Earth. The data shows how the energy from the sun oscillates smoothly throughout the year as the seasons change. In fact, the amount of solar radiation hitting a given spot at any time should be the same year after year because, in terms of human time scales, the sun shines pretty consistently. The NSRD measures changes in the amount of radiation, which are usually caused by changes in the atmosphere such as air pollution. Therefore, these data can be used to measure air quality.

What about somewhere like Miami, where it's much hotter than DC in the summertime and pretty warm year-round? Is there still a change? Yes, yes, there is. NSRD measurements show that the average solar radiation in Miami is higher than in DC and, consequently, the average temperature is higher as well. In addition to the Earth's tilt, an area's position on the Earth matters, too, in terms of average seasonal temperature swings. The angle at which the sun's radiation hits the top part of the globe (higher latitudes) fluctuates significantly throughout the year, whereas the angle of light hitting the center part of the globe (lower latitudes, near the equator) is more consistent. The sunlight Miami receives comes from roughly the same direction throughout the year, but DC gets more direct light when Earth is tilted toward the sun. It's just like the tilted flashlight. Try the flashlight demo on a globe and you'll get the picture.

Unfortunately, I couldn't find a good graph measuring the solar intensity in Dorne, but I think it's safe to assume Dorne's climate is pretty close to that of Miami and Winterfell's climate is more like northern Canada's. The climate of King's Landing is probably similar to that of DC. The seasons most likely change in about the same way, though on a much different scale. And we all know that winter hits Canada much earlier and more dramatically than it does Miami. So, now that we know how seasons work on Earth, we can start looking at how seasons might work on the planet that is home to Westeros.

VERY ELLIPTICAL ORBIT

Even though Earth's seasons are caused by the tilt of its axis and not the change in its distance from the sun due to its elliptical orbit, would it be possible for a planet to have unpredictable seasons because of its orbit? Well, not really. What drives the story in *Game of Thrones* is that no one knows how long the seasons will be. The maesters have been charting seasons for roughly 1,000 years and still can only discern rough patterns, such as a long summer being a good indicator of a long winter. Orbits are very regular, as we know from Kepler. I'm going to guess that after 1,000 years of observing, they would have determined the seasonal variations by now if the seasons were predictably regular on a time scale much smaller than 1,000 years. It certainly took Galileo way less time to figure it out. That being said, the seasons may be caused by a combination of the planet's orbit and something else. First, let's examine what would happen if seasons were caused by orbit alone. What type of difference would there need to be between aphelion and perihelion? Would it be possible for a planet to have an orbit that big?

The average temperature of Earth is about 57°F (14°C), and the average distance from the sun is 149,597,871 km. To reach a temperature of 35°F (4°C)—a common occurrence during a DC winter—you would need to change the temperature by 22°F (10°C), or by roughly 3.5%. Those of you with quick math skills are probably asking how I got 3.5% when your math says it should be 40%. When it comes to temperature, however, looking at relative measurements (such as percentages) has to be done with respect to some definitive zero point. For temperature, the zero point isn't 0°F or 0°C because it's possible to have negative temperatures on both of those scales. The zero point of temperature is zero on the Kelvin scale—0 K, or *absolute zero*. At absolute zero, all molecular motion stops. If 0°C is 273 K, then, we're actually measuring the difference between 287 K and 277K—a decrease of 3.5%. Since the decrease in temperature is equal to the square of the distance, a 3.5% change in temperature would indicate a change in distance of approximately 1.9%. The average distance between Earth and our sun is 149,597,871 km, so 1.9% of that is about 2,900,000 km. At the aphelion, the planet would need to be 2,900,000 km farther from the sun than the average. In the summer, this is reversed and the planet would need to be 2,900,000 km closer to the sun than average at the perihelion. The difference between aphelion and perihelion of such a planet would have to be almost 5,800,000 km. That's about a million kilometers more than the difference between the two points in Earth's orbit. Are there any planets that have that big a difference? Yes—quite a few, actually; five in our solar system alone. Pluto has the greatest difference between its perihelion and aphelion—about 3 billion kilometers. That means its closest point and its farthest point differ by more than 20 times the average distance between the Earth and the sun. Mercury, Mars, Saturn, and Uranus all have differences greater than what would be needed to cause winter and summer. The difference between the perihelion and aphelion of both Mercury and Saturn is roughly 27,000,000 km, which is pretty close to what would be needed to cause winter and summer on our mythical planet. Suffice it to say, the fact that a planet has a highly elliptical orbit could be a possible explanation of seasons in Westeros, but the change in those seasons would be predictably periodic, and I think we can assume the maesters would have been able to figure this one out by now. Just because they are too much in their own heads to figure out that the Night King exists doesn't mean they couldn't figure out that seasons are pretty regular. There has to be another explanation.[5]

MOVING AXIS

I mentioned earlier that the Earth's axis is tilted by roughly 23.5°, and that this tilt causes seasons. This was mostly, but not totally, correct. Much like that old Einstein joke, it's not right, but it's close enough for all practical purposes. Right now, Earth is tilted at about 23.5°, but that tilt is changing ever so slightly every year. And it doesn't just change in one way—it changes in two different ways: it rotates (although the physics-y term is *precesses*) and it wobbles just like a spinning top. Lucky for us, these changes take place over many thousands of years, so we don't really notice them. The gravitational pull of the sun and even the moon affects how the Earth moves. The physics of gravitational interactions among three different objects is extremely complicated and is not something that can be solved easily. In fact, the scientists studying the interactions between the sun, Earth, and the moon got so fed up that they gave the problem its own name—the "Three-Body Problem"—and it is still vexing to this day that there is no actual solution. That's right—we can detect gravitational waves from colliding black holes, but not even Stephen Hawking could write out an equation that would give you the position and velocity of Earth, the moon, and the sun at any moment in time given certain starting conditions. Worse yet, if something were to happen to the moon, the delicate balance of all of the gravitational pulls would be thrown off, potentially causing extreme changes in climate and seasons. It seems as if we got pretty lucky with the stability of the Earth's orbit and its interactions with the moon. It's possible, however, that the Westerosi are not as lucky.

First, let's talk about the precession of the Earth's axis. This is a lot like a nonwobbly top that wasn't spun exactly up and down. As the top (or planet) spins, the axis around which it spins slowly rotates around the line perpendicular to the floor it's spinning on. The top of the axis would trace out an imaginary circle in the air. In fact, astronomers first figured out that the axis was moving because it looked like the stars were moving in an arc in the sky. Turns out it was us moving, not the stars. This movement of the axis of rotation is called *precession*, and Earth does exactly the same thing—except very, very slowly. It takes about 20,000 years to make one full precession. What precession means in terms of seasons is that the seasons don't always fall at exactly the same time of year. If there were no precession, the tilt of the axis would be in the same spot at the same point in Earth's orbit. It would always point toward the sun at one point in its trip around the ellipse—the same point in every revolution—and away from the sun at the halfway point in its journey. But because of precession, this isn't the case. The point in the orbit when Earth points toward the sun (meaning summer) changes slightly every year. Because it is such a tiny change, this isn't something we are going to notice in our lifetimes or even on the scale of the maesters' recordings. It's possible, though, that the axis of the mythical planet in *Game of Thrones* precesses much faster than Earth's axis. This change in tilt direction could be responsible for the strange seasons in Westeros—but, again, the seasons would change in a predictable way, as seen in a very elliptical orbit. If the maesters had been charting the seasons for thousands of years, this is a pattern they should have been able to see by now. They'd know when winter was coming.

Precession isn't the only way Earth's axis moves, however. Earth's axis doesn't stay fixed at 23.5° and precess around; the actual angle changes as well. It wobbles. The angle changes only in minuscule increments—about 1.3° over the course of 21,000 years. If you've read the section on why we have seasons, you can extrapolate from that information that the change in tilt will affect the seasons somewhat. We are lucky that this change is very, very small. In total, our axis shifts by only 10% of its average angle, give or take. Percentagewise, the axis is about as stable as our orbit.

Why it changes is a slightly more complicated situation. In 1993, a French research group—Jacques Laskar, Frédéric Joutel, and Philippe Robutel—simultaneously published two papers in *Nature* that discussed the moon's effect on Earth's axis.[6] (Most scientists would kill for one paper in *Nature*; two in one issue is just overkill.) I said earlier that it wasn't possible to write an equation to solve for Earth's position at any given time with certain starting conditions. This is true, but we can use computers to make a very close approximation. There are equations that describe the motion on Earth in a complicated way, but they aren't equations that we can sit down and solve. In big math-y terms, they are partial differential equations with no exact solution. It's just as scary as it sounds. Computers, however, can employ methods that use numbers that are close to a solution but not actually a solution, which gets them pretty close to an explanation of how the moon and sun affect the movement of Earth. Laskar, Joutel, and Robutel looked at the equations for the motion of Earth's axis and asked what would happen if the axis was tilted at something other than 23.3°. What they found was pretty surprising.

According to their simulations, if it weren't for the moon, Earth's axis would be thrown into chaos. No, really—it would be scientifically chaotic. Much like babydoll dresses and chokers, nonlinear dynamics and chaos were all the rage in physics in the '90s. In a system, if you move one thing a little bit, something else usually moves a little bit in response. A bigger initial movement results in a bigger change. In nonlinear dynamics, all bets are off. In a nonlinear system, a small movement can cause a big change—or maybe no change at all. It's a pretty neat branch of physics and is one of the most commonly invoked "science-y" terms in movies (think *Jurassic Park* and Jeff Goldblum's incessant whining about the butterfly effect), right behind quantum mechanics, relativity, and wormholes. In nonlinear dynamics, chaos can be characterized mathematically by recording a lot of data on the state of the system and then using fancy statistics. Something that is chaotic may appear random but will have underlying order. There are repeated patterns, feedback loops, and quasiperiodicity, which means that something may look like it is periodic, but it just isn't quite there. For example, something may follow one cycle a few times before moving to another that is very similar but not quite the same. Sounds a lot like the seasons in Westeros, huh? The problem with the Westerosi maesters, however, is that they don't have enough data (much less computers!) to even guess at a pattern that is not periodic.

Laskar, Joutel, and Robutel found that if they set up a system with the sun, the moon, and Earth tilted between 0° and 60°, the tilt of the Earth would remain pretty stable, changing by only about 10% and moving in a predictable manner. When they increased Earth's tilt beyond 60°, however, they found that the axis entered a "chaotic region," meaning that not only was the tilt changing quite a bit, it was also doing it in a quasiperiodic or even a completely unpredictable fashion. Things got really interesting when they took away the moon altogether. With no moon, any axial tilt falls into the chaotic region, meaning Earth's axis would not be stable at all. The moon stabilizes the dynamics of our axis and stops it from moving chaotically. The moon, therefore, is responsible for more than the tides; its stability is why we have regular seasons.

Milankovitch Cycles

There are clearly a lot of things going on with a planet's orbit that could contribute to its seasons: the orbit shape, the axial tilt, and the axial precession. All of these variables affect a planet's climate and can potentially lead to an unpredictable pattern of seasons. What do seasons look like when you add up all these different changes? What if a planet with an aphelion that drastically differed from its perihelion also spun on a wobbly, quickly precessing axis? The sum of these changes (in addition to some much smaller changes that I haven't addressed) determines the planet's Milankovitch cycle.

In the 1920s, the Serbian geophysicist and astronomer Milutin Milanković wanted to identify the effects of Earth's changing orbit on the planet's climate. He studied the different aspects of Earth's orbit and how they changed, and predicted how Earth's climate might change as a result. Because Earth's orbit, tilt, and precession occur so slowly, he reasoned that the change in climate would occur over the course of thousands of years. He assumed that the amount of solar radiation received by Earth was responsible for large-scale climate changes such as ice ages.

Milanković's research drew on the work of two earlier astronomers, Joseph Adhémar and James Croll. Adhémar was the first to suggest that ice ages were caused by Earth's orbit, and he went on to theorize that ice ages were caused solely by precession of Earth's axis. Croll built on his work, proposing that the shape of Earth's orbit was a causal factor. By his predictions, however, ice ages should have stopped about 80,000 years ago, and, well . . . that's not what the data say. Milanković went further than either of these astronomers, examining all of the variations in Earth's motion and combining these discrete observations to understand how the solar radiation received by Earth would change over the course of millions of years. These long-term changes are now called *Milankovitch cycles*.

In 1976, *Science* published a paper by James D. Hayes, John Imbrie, and Nicholas Shackleton that supported Milanković's theory. They looked at sediment cores taken from a point in the ocean floor between Africa, Antarctica, and Australia and determined that the rate of sediment deposition was constant, thereby providing a time line of events in geological history. By calculating the concentrations of microorganisms in different parts of the core, they were able to estimate the average water temperature when each layer of sediment was deposited. From there, they could determine points in time where Earth had experienced an ice age. Milanković's theory uses calculations of Earth's temperature based on its movement to accurately predict the onset of these ice ages. Since 1976, Milanković's theory of the origin of ice ages has been the prevailing theory.[7]

A distinction needs to be made between climate and weather, and now is as good a time to do it as any. Milanković looked at climate, not weather. *Climate* refers to the long-term changes in temperature and precipitation averages. *Weather* is the term for those changes on a day-to-day basis. When talking about seasons in Westeros, I'm talking about weather, not climate. In his research, Milanković was looking at climate rather than weather, so at first glance his work seems only mildly relevant given that we're already operating under the assumption that it's possible to speed up changes in a planet's tilt, elliptical orbit, and axial precession to a significant degree. Those processes, however, happen very slowly on Earth, so that's what Milanković was studying. If the changes that happen in the long term on Earth occur in the short term in Westeros, it's possible that Milankovitch cycles could explain seasons. Milanković's time scale was imposed because that is the time scale of Earth, but there's no reason it couldn't be sped up under different cosmic conditions such as those existing in Westeros.

Two Suns?

On April Fools' Day several years ago, some enterprising fantasy-loving grad students decided to investigate what may be causing seasons in Westeros. Although they approached it differently than I have so far, they also came up with an explanation for Westeros's variable seasons. The Maesters of their Citadel were not as excited about their theories as I am. Luckily, one member of the team, Veselin Kostov, answered my raven, and I was able to talk with him about their theory and simulations. He had been working on simulating the orbit that a planet might have around a binary star system. He noticed that the surface temperature of a planet orbiting a binary star system fluctuates much as the seasons did in a book he happened to be reading at the time. Being that it was April Fools' Day and he had the program to do it, he decided to simulate the planet of Westeros as if it were orbiting a binary star system. He dismissed axial tilt as an explanation due to the stabilizing force of the moon. (I think he was too quick to make this assumption, but I'll talk about that in the next section.) He also dismissed the argument for a highly elliptical orbit as the cause of the irregular seasonal pattern, but for different reasons than I have—Kostov argues that it would lead to seasons so long that the planet could not have sustained life. So, with two different arguments against it, I guess we can throw out the elliptical orbit explanation.

Kostov's assumption was that Westeros's planet orbited a solar system that revolved around not one sun but two. (I am sure the group's analysis would also apply to Luke Skywalker's home planet of Tatooine.) Using numerical solutions, Kostov showed what would happen over many years under a certain set of initial conditions. He assumed there was no axial tilt and that the surface temperature was based on the combined heat of the two stars. This involved a pretty complicated equation that calculates surface temperature based on the planet's distance from the suns. He used a computer to solve this equation, assuming that the distance from the suns was in accordance with the complicated three-body problem I described earlier in the chapter and that a Westerosi year was approximately 700 days long. It wasn't explained in the paper where this number came from, so I asked Kostov, who said it was the average length of an inhabitable planet's orbit around two stars. His results were pretty close to what seasons in Westeros look like. The temperature is quasiperiodic—so it's almost predictable, but not quite—and changes on the order of a few Westerosi years or so (figure 1.3). This seems like a pretty good explanation, but there are a few problems with it, some of which I've already mentioned.

First, Westeros seems to have only one sun. My brain is not flexible enough to do the mental gymnastics needed to overcome this point. I simply cannot figure out a physically possible way to hide a second sun amid the stunning visuals and attention to detail in both the books and

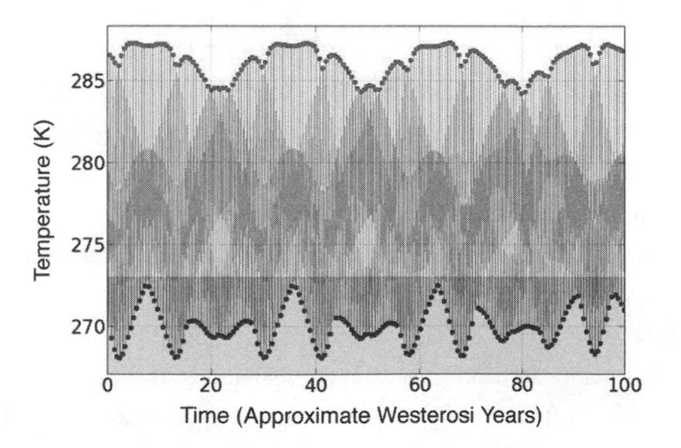

FIGURE 1.3
Surface temperature of a planet orbiting two stars with a 750-day orbit. The bottom line represents winter and the top line represents summer.[8]

the TV series. I got a chance to ask Kostov this question, and he said it wasn't something he had thought about; if pressed, however, he said it might be a combination of a star and a black hole. Changing one sun to a black hole would certainly affect the temperature data, so more simulations would be needed. Secondly, defining the length of a Westerosi year is more complicated than these maesters let on. They looked only at the distance between the stars and the planet; they didn't plot the actual orbit. Because one year equals one orbit around the sun(s), there is still the issue of determining the length of a year. It matters less in this case, because without axial tilt there would be no real way for those on the planet to figure out how long a year is. But that makes it all the more complicated. The inhabitants of Westeros talk about summer lasting 10 years, but with no markers to indicate the passing of each year, how would they know this? Without astronomical markers, how were they able to determine the number of days in a year those many thousands of years ago? I think Kostov and his team did a great job of presenting a possible explanation for Westeros's seasons, especially given their lack of funding and the brief amount of time they had. I don't think this is the best explanation, however.

SO, WHAT ABOUT WESTEROS?

From all the science you just read, I hope it came through that it is really easy for a planet to have very strange seasons. Things can swing wildly for a number of different reasons. The reason why seasons in Westeros are hard to explain is not that seasons can't be that long but that seasons are rarely that unpredictably wonky. To make it more complicated, the seasons aren't just completely irregular; they seem to follow some patterns, making them quasiperiodic, as discussed in the section about chaos. It is said that a long summer is followed by a long winter, a belief so commonly accepted by the Westerosi that people are bracing for an exceptionally long winter after the 10-year-long summer, even without the Night King's efforts to make that long winter eternal. The seasons also need to change on the order of years, not months as they do on Earth. This further constrains the possible explanations. Because seasons are governed by a planet spinning, orbiting, tilting, and precessing, they are (in most cases) very, very regular, and in the case of Earth, they are much shorter than those in Westeros. The one exception to seasons' regularity is axial tilt, which I'll talk about shortly. Strange season length may be a great plot device, but it's not that easy to explain scientifically. I think, however, that there are ultimately two reasonable explanations, and I have a favorite. I'm going to dismiss the two-suns argument altogether because it doesn't fit in with the mythology at all—this is the Known World, not Tatooine. Someone would have mentioned seeing two suns in the sky at least once in the 5,700 pages of *A Song and Ice and Fire* currently in print. I am going to try and keep all my scientific explanations within the world GRRM expertly built. I applaud the grad students' use of their free time, but I wish they had spent more of it reading the books.

The first reasonable explanation—not my favorite one—is that the planet has a very complicated Milankovitch cycle—complicated enough that, even after thousands of years, they can just barely make out a pattern. This is certainly a reasonable explanation. If the orbit were elliptical enough, the axis wobbled fast enough, and the axis precessed quickly enough, this could result not only in season changes but in ones erratic enough that the maesters wouldn't see a real pattern within their historical record. Long summers might not be followed by long winters, and it's possible that the seasons would change too quickly to fit in the *Game of Thrones* universe. A complicated Milankovitch cycle might also appear almost, but not quite, periodic. There might be some regularity, but not enough to be reliable. Normally, this is where I would do the math to show how it might work. That's difficult in this case, though, because there are three things changing at the same time, and a Milankovitch cycle adds all three of those up. I would have to start with a Milankovitch cycle that fits the requirements and then mess around with three variables to make it work. This is like starting with the number 20 and figuring out some of the many ways to add numbers together to get there, only much more complicated. The important part in this discussion is whether or not it's possible to get to 20, or rather to quasiperiodic seasons that change on the order of years—and with so many variables, it certainly is.

The second possibility, and by far my favorite, is that the axial tilt is chaotic. We know the moon is responsible for the stability of the Earth's axis and thus for the regular and short seasons. We also know that if there were no moon, or even if Earth's axial tilt were greater than 60°, Earth's axial tilt would become erratic, chaotic, or quasiperiodic. What if the Westerosi planet lost a moon somehow? Without that stability, the planet's axial tilt could be sent into chaos. Whether intentionally or unintentionally, GRRM wrote this into the legends of the free city of Lys in Essos:

> "He told me the moon was an egg, Khaleesi," the Lysene girl said. "Once there were two moons in the sky, but one wandered too close to the sun and cracked from the heat. A thousand thousand dragons poured forth and drank the fire of the sun. That is why dragons breathe flame. One day the other moon will kiss the sun too, and then it will crack, and the dragons will return."

Though Laskar and Robutel looked at the effects of a single moon on Earth's axis, we can assume that two moons would add more complexity. If our one moon were to wander too close to the sun and crack, our planet's axis would literally be sent into chaos. With the delicate balance of two moons acting on a planet in a way that was stable enough to evolve intelligent life, losing one would have a serious effect. The immediate disruption of the balance could easily send the planet's axis into a very wobbly state and create the erratic seasons.

There is another reason I think this is the best explanation. Something I haven't talked about yet (which many of you readers have surely figured out) is that having a season that lasts more than a year is really, really difficult. On Earth, because of our axis' tilt, we go through all of our seasons in one year. It has to happen that way. One year is defined as one trip around the sun. On any planet with a mostly stable axis, it will have to, by definition of the term *year*, experience all seasons in one year, or one trip around the sun. There's no way of getting around that. If the axis is changing, however, that's no longer the case. With a changing axis, all bets are off on how long a season can last. It's the only way you can have a planet that has a season that lasts more than one year. It can certainly be intensified with a complicated Milankovitch cycle, but at the end of the day (or year, or summer, or long winter), the only way you can have a 10-year-long summer is to have an axis that tilts in a very specific way as it goes around the sun. For 10 years, the axis would have to be tilted toward the sun, and then it would have to be tilted away for several years.

The axis would have to be almost constantly changing its direction, on the order of days, to make this happen. As seen in the discussion of the chaos planets can be thrown into when moons go away, this is certainly a possibility, but it would be a very unusual outcome.

Do I, as a scientist, think it's possible for a planet that has seasons like those of Westeros to exist? I think it certainly is. The explanations exist to describe how a planet could behave like that. There is a legend in the world of *Game of Thrones* that explains how it could happen. There are so many planets in the universe that it's possible one could be in the "Goldilocks zone" and develop life while also having funky seasons. Do I think it's all that likely? Nope. But I'm also (or I like to think I am, anyway) an intelligent being writing this chapter on an iPad developed with insanely complicated technology while sitting in a coffee shop that can produce a huge variety of amazing beverages. All of that seems rather unlikely as well, considering everything on Earth started as goo. So, can I rule out a planet that has a perfectly chaotic axis, at a perfect distance from a single star, in an orbit that is just right, with a moon that makes it *just* unstable enough? Not in the slightest.

2

AND NOW MY WATCH BEGINS

THE SCIENCE OF AN ICE WALL

You could see it from miles off, a pale blue line across the northern horizon, stretching away to the east and west and vanishing in the far distance, immense and unbroken. This is the end of the world, *it seemed to say.*

—Jon Snow on the Wall, *A Game of Thrones*

Ice. Solid water. It seems like this is a pretty straightforward concept. Ice is something everyone is pretty comfortable with, whether it's floating in your Diet Dr. Pepper or you're gliding across it on ice skates. But ice is way more than that. It is possibly one of the coolest (haha, get it? *Coolest*?) solids on the planet. As I'm sure you know, water sustains life on Earth. One of the key attributes that helps it in this endeavor is that its solid form, ice, is less dense than its liquid form. Fish wouldn't survive long if their lakes and streams froze from the bottom up. In Westeros, ice serves yet another lifesaving purpose: it keeps out the White Walkers. I know that magic was woven into the Wall when Bran the Builder built it thousands of years before the plot of *Game of Thrones begins*, but I'm going to approach understanding ice and the Wall assuming only the magic of some physics and see what I come up with. Could we build a giant ice wall? Would it be worth it? What would it look like? Could it be destroyed by a magic fire- (or, I don't know, a cold- and ice-) breathing dragon? How hard would it be to climb such a wall? Before launching into the science, it's worth laying out some specifics. According to GRRM, the Wall is 300 miles long and 700 feet high, and in most places, the top is wide enough for 12 mounted knights to ride abreast, on average. It was built over many years, starting small and slowly growing. It often weeps but never fully melts, and the wildlings who have climbed it can tell you that it's possible for large chunks to break off. Yet it has stood for 8,000 years.

WHAT IS ICE AND HOW DOES IT WORK?

As with seasons, this seems a weird question to ask. Everyone knows what ice is. It generally comes out of the freezer and often goes in a Manhattan after a long day at the office. It's solid water. How complicated can that be? It turns out that ice can be quite complicated. There's not just one type of ice crystal, but 17 different types that we know about so far. If that's not enough, there are three different types of amorphous ice. The best-known property of ice, as everyone who has ever left a beer in the freezer for too long knows, is that it expands when it freezes. And it doesn't just expand a little bit—it expands a lot, generally enough to break glass and crack rock and even metal. It's also responsible for the post-winter potholes in the colder regions of the United States. The fact that it expands when it freezes is why ice is less dense than water. There are many chemists and biologists who would say this property of water is why we have life on Earth at all. Think about it. If ice were denser than water, then instead of lakes having a coating of ice, the frozen water would fall to the bottom of the lake. Bodies of water would freeze from the bottom up. That's not real great for fish, algae, and other aquatic organisms. Having a layer of ice on top of the water also helps insulate what's underneath. Fish don't freeze. Water is the only material that is found naturally in all three phases on Earth. It is also the only substance that has a different name for every state. This is rather annoying when trying to write this chapter. According to GRRM, ice can also be a fabulous defense mechanism.

Before talking too much about using ice as a building material, it's worth talking about the fundamental structure of ice. As I said before, there are 17 different types of ice crystals—or at least right now there are. Scientists seem to keep creating more and more; if you search for the number of phases of ice, you'll find anywhere from 11 to 17. Recently, a paper was published in the *Journal of Chemical Physics* that suggested there could be as many as 300 phases of ice. That's all theoretical, however, so it doesn't quite count yet. The 17 phases I'm talking about have all been found in an experimental environment, even if only for a fraction of a second.

For the most part, we are used to what's called Ice I_h, which is pronounced "ice one h." (I had to look that up.) The "h" in Ice I_h stands for "hexagonal." In this phase of ice, the oxygen atoms sit in a hexagonal lattice. It looks quite a bit like a honeycomb. Water molecules are composed of one oxygen atom and two hydrogen atoms that are attached to each other because they share some electrons—two electrons, to be exact. Much like a class of questionably parented kindergarteners, they don't share evenly. Oxygen is a bit greedy and tends to take the electrons. A water molecule as a whole doesn't have any charge, but the hydrogen end is a bit positive because it is short the electrons it "shared" with oxygen. The oxygen is a bit negative because it pulls the two extra electrons in its direction. In water, there is enough heat for the molecules to move around wherever they want. But opposites attract, and the negative oxygen likes to be near the positive ends of other water molecules. They really cuddle up, making water fairly dense. These types of bonds are called *hydrogen bonds*. They're the reason why water tends to stick to itself and flow in a tight stream. But when water cools down, the water molecules get sluggish and can't move around as much. They eventually sit down in an ordered hexagonal lattice and become Ice I_h. In a honeycomb shape, however, the molecules are spread out quite a bit more than they are in the cuddle party that is liquid water. Because there are fewer molecules in a given volume of ice, it's less dense than water.

Normal hexagonal ice makes up pretty much all of the ice you come into contact with. In talking about Westeros's Wall, it's really the only phase that matters. But there is one other naturally occurring phase of ice, and it's worth talking about. The difference between different phases of ice is the crystal structure. The only other naturally occurring phase of ice is Ice I_c, pronounced "ice one c" (or the way more fun-sounding "ice icy"). The "c" stands for cubic. Once the temperature of water drops below −50°C, the crystal structure changes from something resembling a honeycomb to a crystal structure similar to that of a diamond. This type of ice occasionally appears in the upper atmosphere. The other phases of ice occur at various temperature and pressure combinations. The phases of ice, water, and vapor are summed up in a *phase diagram*. This is a handy little chart that shows the form water will take at a specific temperature and pressure. As I said before, the only ice that really matters in discussions of Westeros is I_h, so from now on, when I say "ice," let's just assume that's what I'm talking about (figure 2.1).

Something worth noting is that there is a part of the phase diagram, between approximately 1 and 200 megapascals (MPa) of pressure, where the melting temperature of ice decreases as the pressure increases. It's hard to see in the previous phase diagram, so figure 2.2 gives you a closer look.

To be clear, 200 MPa is a large amount of pressure—about 100,000 times the air pressure on the Earth's surface—but it's worth keeping in mind that this is in the context of building with ice. 200 MPa is equal to about 200 million newtons (N) of force. The average elephant exerts a force of (or weighs) about 20,000 N, so for ice to reach the point where

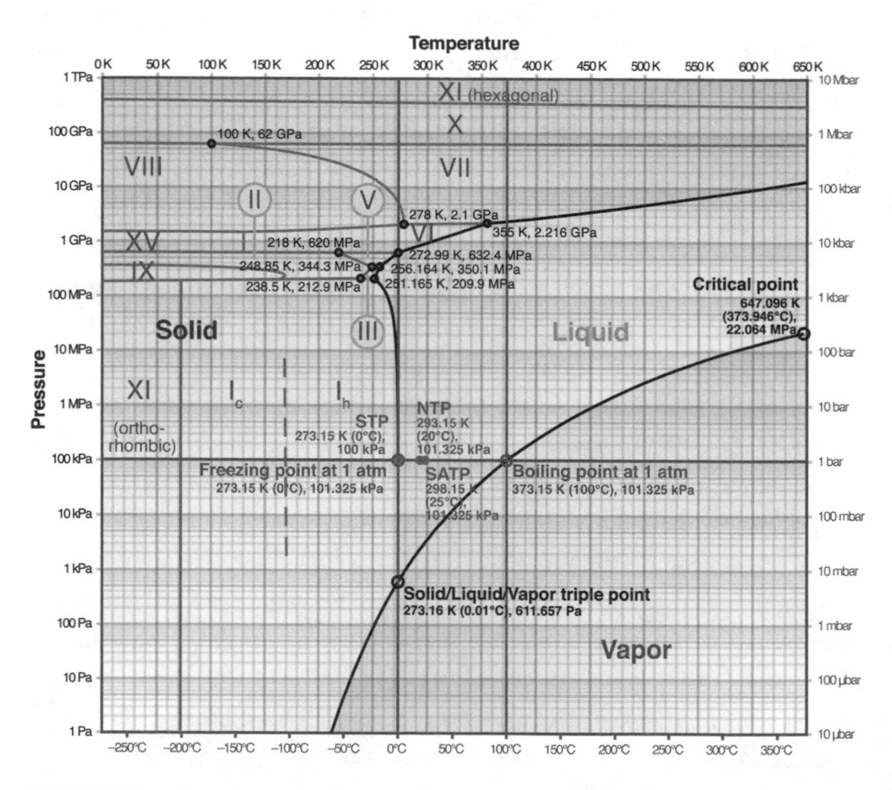

FIGURE 2.1
Full phase diagram of ice

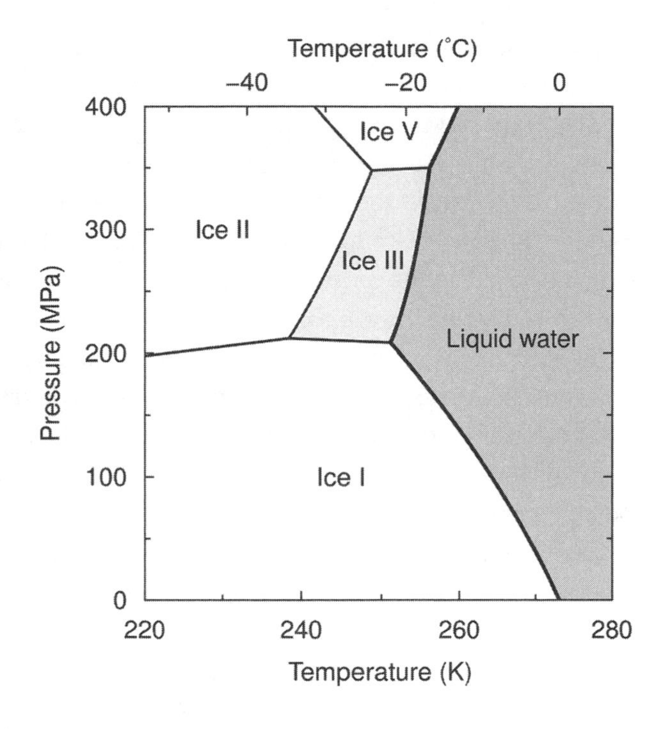

FIGURE 2.2
Useful phase diagram of ice

pressure would lower the melting temperature, you would need 50–1,000 elephants standing on a single square meter. The more elephants, the lower the temperature at which ice will melt. For a long time, this property to melt under pressure is what people thought was responsible for ice being slippery. If you listen to most kids' explanations of how ice skates work, you'll think that the pressure from the weight of a person on such a small skate blade causes enough pressure for the ice to melt. Unfortunately, this explanation is about as accurate as the ones about seasons those Harvard students came up with in chapter 1. People just aren't heavy enough to cause the ice to melt under pressure. Figuring this out left scientists with a bit of a problem because for a long time there was really no good explanation of why ice was slippery—anyone who has ever wiped out on a patch of ice knows that it is, but no one could explain *why*. It was theorized that ice was slippery because the molecules on the very top couldn't bond with the rest of a block of ice, so they were acting like a liquid and making the surface of the ice slippery. This was a great idea; however, with the advent of atomic force microscopes, researchers were able to see that this wasn't the case. What actually makes ice slippery is the interaction between the ice and whatever it is sliding on.[1] The friction causes the top layer of ice to melt just a little bit, and the water is what makes it slippery. The slipperiness of ice depends on the temperature and how fast you are moving across it.

So far, this chapter has covered how ice freezes and what it looks once it's frozen. But considering that there are seasons in Westeros and that the Wall isn't always the same temperature, it's worth talking about how and when ice melts. From the phase diagram above, we know that ice melts at different temperatures under different pressures. Temperature is a measure of how fast molecules are wiggling. The faster they are moving, the higher the temperature. As the molecules sit in their hexagonal ice crystal form, they are still wiggling a little bit depending on how warm they are. As they warm up, they wiggle more and more. Eventually, they wiggle so much that they amass enough energy to break the bonds of the crystal and transition from a solid form to a liquid form, which makes it easier for them to move around. The temperature that triggers this change is called the melting or freezing temperature. It's the temperature at which water turns to ice and ice to water.

Now, you're probably thinking, "Hey, how can ice and water exist at the same temperature? How come sometimes it melts at that temperature and how come sometimes it freezes?" Those changes are based on the available energy around whatever you are trying to melt or freeze. If ice is surrounded by something warmer, the heat from the surrounding stuff will make the water molecules in the ice shake more and more until they break apart and the ice becomes water. An important thing to note here is that the energy needed to melt something isn't just the energy needed to warm it up. It's the energy needed to break the bonds between the molecules plus the energy required to warm up the ice to the melting point. That extra energy is called the *latent heat of fusion*, and for ice it's about 334 kJ/kg. To give you an idea of what that number means, the average space heater gives off 1 kJ per second, so it would take the energy in that space heater about five and a half minutes to give off enough energy to melt a 10 cm^3 block of ice. It takes tons more energy to melt ice than it does to warm ice from say, –50°C to –20°C. The surrounding stuff (the source of the energy that is warming the ice) loses energy as the ice warms up and melts. It loses a lot of energy in melting ice. That's how ice keeps your Diet Dr. Pepper cold. The heat from the soda is used to melt the ice so the surrounding soda cools off. This is also why whiskey stones are a total sham. Seriously, I can't stress this enough: don't buy whiskey stones. If you want to keep your drink cold without watering it down, get yourself some water-filled plastic ice cubes—they're 80% less stylish, but 100% more useful, and here's why: A 0°C whiskey stone and a 0°C ice cube will cool your drink very differently. The stone will take away enough energy to raise its temperature to the temperature of the whiskey. An ice cube will take away the same amount of energy as well as the energy needed to break the molecular bonds, which melts the ice. It will keep your drink a lot colder. When water is surrounded by a colder substance, the opposite happens. The molecules slow down and cuddle up with their neighbors to form crystals. When ice is under pressure, the freezing/melting temperature changes based on how much pressure is exerted on the ice.

Like all materials, ice has a point at which it will crack. This is determined by how the ice crystals are organized and how regular they are, as well as the strength of the bonds that keep the crystal structure together. Because the hydrogen bonds that hold the water molecules in place aren't super strong, ice generally isn't winning any strength awards.[2] Like anything, ice isn't perfect. There are small defects everywhere. Within each cell of the honeycomb structure, the molecules may not be sitting evenly. This is called a point defect. There may be strips of well-organized crystals sitting next to some that are off by a bit. This is called a dislocation line. If those two issues weren't enough, sometimes one layer doesn't lie flat on top of the next layer. This is called a planar defect. These defects cause cracks because it's much easier to make something slip or move around when it isn't firmly in place. In the case of ice, there tend to be a lot of defects and thus a lot of ways for cracks to form. This makes ice brittle. These tiny cracks and defects are important on a small scale when dealing with things like ice cubes, but on a large scale under a lot of pressure, they can lead to some really interesting ice structure dynamics. I'll talk about that a bit later.

During World War II, a group was studying how well ice could hold up to bullets and other artillery for military purposes. What they found was that ice was weak, but it wasn't constantly weak. Some blocks of ice would shatter under light pressure, and some could take much more force. However, the ice wouldn't break under the same amount of pressure every time. This was because different blocks of ice had varying amounts and types of defects. There are ways to get around this, and the engineers and scientists of WWII spent a lot of time on this problem.

THE AMAZING PYKRETE

Bran the Builder wasn't the only one to think of making large-scale structures out of ice—engineers in World War II took a crack at it, too. Ice may be good at defending against fire, but it cracks easily and doesn't hold up well under pressure. In addition, it never quite fails in the same way. Even with low-quality materials, if engineers know how they are going to fail, it's possible for them to devise a work-around. By contrast, there's no specific set of conditions that cause ice to fail; rather, it fails under a wide range of conditions. This made engineers really throw up their hands. It is possible to make ice stronger and more consistent in failure, but it needs some help getting to that point. When ice is mixed with other materials, such as gravel or sand—or, better yet, wood pulp or Kevlar—it can be quite strong.

World War II necessitated a lot of creative engineering. Money was short, and the war was long. Scientists were turning to alternative materials to make crucial matériel such as planes and ships. The Allied army wanted complete aerial coverage of the North Atlantic. Aviation technology wasn't advanced enough to allow them to do this from land, so they decided they needed to do it from the water. They wanted to build a system of aircraft carriers that could sit in the ocean and allow planes to land and refuel. Essentially, they were trying to build a small archipelago. They also wanted it built cheaply and with easily available materials. Oh, and it had to be unsinkable by German U-boats—ya know, no biggie. To do this, they turned to ice. In 1942, Geoffrey Pyke suggested hollowing out an iceberg for just this purpose. It may sound crazy, but they didn't need the aircraft carrier indefinitely; it would only need to carry planes until the military could establish land bases, and a hollow iceberg certainly couldn't be sunk. Churchill thought the idea merited further exploration. Unfortunately, ice floes were too thin and icebergs didn't rise high enough above the waterline to make the so-called bergship. After reviewing current and conflicting research about ice strength and doing some of their own research, the Allies realized ice was just not going to cut it as a building material. They would have scrapped the project if it weren't for a paper by Herman Mark and his assistant Walter P. Hohenstein. Following a hunch, the two had embedded some wood pulp in water and froze it. They found that the new material was much stronger than ice. This discovery revived the bergship project and sparked new research on "pykrete" (named after Pyke).

Researchers began testing the strength of pykrete in comparison to ice and to the required strength of a ship. They experimented with different concentrations and kinds of wood pulp. Pykrete containing as little as 4% wood pulp proved vastly superior to plain ice. The rocks or other materials mixed in with the ice helped stop cracks from propagating. Normally, a crack will continue along fracture planes and defects until it either runs out of energy or reaches an edge, at which point a chunk of ice will break away. Materials such as wood pulp act as roadblocks, strengthening the ice by preventing cracks from advancing. As a crack progresses, it will eventually hit a relatively large object, be it a stone or a piece of wood, which will stop it in its tracks. Furthermore, even if the pykrete were to crack, the added material would help it fail "elegantly," or in a predictable fashion. It wasn't perfect, by any means, but where normal ice beams would break under pressures of 490–3432 kPa, pykrete could withstand pressures of roughly 3678–6129 kPa. On average, pykrete was about twice as strong as ice under compression and about five times as strong under tension. The researchers found that a concentration of approximately 4% wood pulp was the sweet spot to aim for in order to create the strongest pykrete, with diminishing returns at higher concentrations.

It was also important to examine how bullets and explosions would affect pykrete. A rifle bullet would travel 35 cm into a block of ice, whereas in pykrete at −7°C it would stop at 16 cm. Pykrete held up to bullets about twice as well as ice. A torpedo would lodge itself about 60 cm into the ice and would leave a crater with a radius of 4.5 m, so the Allies planned to make the bergship 9 m thick as a precaution. This resistance to firepower means that even rocks hurled by catapults wouldn't stand a chance against the Wall, assuming that it was made of some Westerosi version of pykrete.

The project was ultimately abandoned, primarily due to advances in aviation technology. The Allies needed to overcome huge technical difficulties for the bergship to become a reality, but the World War II era was a time of making impossible tech work. With the construction of more aircraft carriers and improvements to airplanes that allowed them to fly longer distances before refueling, there was no longer an urgent need for the bergship to create a floating island system in the North Atlantic.[3]

Ice on a Large Scale Is Basically Ketchup

When I first started the research for this chapter, I was convinced that the ice wall would deform because of the melting of ice under pressure. I'd done the physics demos of pressing string into ice blocks, and I'd even taught lessons about how ice skates work. Once I started digging into peer-reviewed research, however, I realized I couldn't have been more wrong—not about whether or not an ice wall would work, but about why an ice wall *won't* work. Then I had a chance to talk with one of the foremost glacier scientists, Dr. Martin Truffer from the University of Alaska, and he quickly set me straight. At the insistence of his graduate students, Dr. Truffer presented work on a theoretical ice wall at the 2017 meeting of the American Geophysical Union, and he's also been quoted in several media outlets discussing the plausibility of an ice wall of that size. He's not a fan of the show, however; he contemplated giving it a try, but had second thoughts after I rattled off some of the other chapter titles in this book. But he didn't need to watch the series to feel confident in telling me to approach the ice wall question by treating ice on that large a scale as a non-Newtonian fluid.

In a structure such as an ice wall or a glacier, the pressure from the structure's own weight causes it to move slowly, or *creep*. Creep happens even at very low temperatures, where ice would not melt on a large scale even under extreme pressure. The majority of creep is caused by dislocations, or small cracks that cause ice crystals to move over each other. Under extreme pressures, it's possible for the crystal structure of ice to be pushed past a limit it's happy with, causing the crystals to deform. Atoms will also move across crystal boundaries. This happens more often at higher temperatures because the atoms have more thermal energy and are moving faster. This means they are easier to push around. The added pressure from the ice structure's bulk directs all of that jiggling energy outward. The ice never melts and thus never leaves the crystal structure; the crystal structure as a whole is subjected to just enough pressure to move around.

For something like glaciers and ice walls, it's most useful to look at how the whole glacier moves instead of how each molecule moves—seeing the forest for the trees and all that. On a large scale, ice acts like a fluid, but a very special kind of fluid. Most fluids have an unchanging *viscosity*, which is a measure of how easily they flow at a given temperature. Alcohol, for example, flows easily even when ice cold, meaning it has a low viscosity, whereas molasses in winter has a high viscosity because it moves far more slowly. In addition, at any given temperature, neither of these fluids' viscosities are affected by how quickly they are made to flow, that is, how hard they are pushed. These types of fluids are called *Newtonian fluids* because they behave in accordance with the physics developed by Isaac Newton, one of the most influential physicists of all time.

There are also *non-Newtonian fluids*, however, and they work a little differently. These types of fluids change their viscosity when you push them, and are known as *shear-thickening* and *shear-thinning* fluids. Shear-thickening fluids stiffen up when they are pushed. If you have kids, or if you just like messing around in the kitchen, you've probably played with the classic example of a shear-thickening fluid: oobleck, a paste composed of cornstarch and water. Shear-thinning fluids, by contrast, decrease in viscosity when they are pushed. The best-known example of this is ketchup. Ketchup's shear-thinning property is why tapping the 57 on a bottle of Heinz makes your fries red. Yes, this really does work—your buddy was right, physics says so. Ketchup is a great shear-thinning fluid; the harder you whack the bottle, the easier it flows. Similarly, ice in large structures behaves like a shear-thinning fluid. The harder you push on it, the more it flows. It doesn't flow very quickly, but it doesn't sit still, either. So, when there is a large quantity of ice under a lot of pressure, it can move around a lot—it creeps.

Ice breaks easily for the same reason it creeps. Creep, to some degree, is merely ice sliding against itself along small fractures in the ice. The ice moves along fracture planes and dislocations bit by bit, causing the whole structure to move slowly over time. When there is a huge mass of ice that isn't easily fully fractured, the fractures cause it to slowly move. Creep is highly dependent on temperature. If you've seen the movie *Frozen*, you'll remember the great scene at the end where Olaf begins to melt, his face slides down, and he attempts to prop it back up. This is creep at a pretty high temperature. A material will creep more quickly as its temperature nears its melting point. When a substance first begins to creep, it is under a lot of pressure and will thus deform quickly. At the same time, however, the strain on the material decreases. This is a lot like *work hardening* of steel, which I'll talk about in chapter 5. As the creeping material moves, it introduces more and more dislocations, which serve to harden the material and stop it from moving as quickly. Gradually, the combination of strain and increased dislocations reaches an equilibrium. At that point, creep slows to a fairly constant rate. It doesn't stop; it's just not changing how quickly it's moving. Eventually, the crystals melt and refreeze in a way that is more conducive to slipping. When that happens, the creep rate increases significantly. These three stages of creep are called primary, secondary, and tertiary. If you look up a specific value for a creep rate, the numbers you'll find most likely refer to secondary, or steady-state, creep. Under really high stress, secondary creep can be so brief that it seems almost as if it were skipped altogether. Ice structures such as glaciers and ice walls are not creeping over smooth ground. There is friction between the ice and the ground, which slows down the creep. The amount of friction turns out to be dependent on how fast the ice is creeping. When you rub your hands together and produce heat from friction, they get warmer the faster you rub them. It's the same with ice: the faster it moves, the warmer it gets. Unlike with your hands, however, ice becomes more slippery as its temperature increases, reducing the friction and thus increasing the rate of creep. Friction is at its highest during the secondary phase of creep. This, like pretty much everything else involved in creep, is highly dependent on temperature. The final shape of the ice wall and its motion over time ultimately depends on how cold it is in the North and when, exactly, winter is coming.[4]

With these characteristics of creep in mind, Truffer set out to model the Wall of Westeros and how it would change in shape over time due to creep. In the case of an ice wall, it's easiest to create a two-dimensional model and measure its height and width over time. If it behaves as it does in the show, the Wall should remain 700 feet high by 300 miles long over thousands of years. To create his model, Truffer turned to a paper by P. Halfar, who measured the dynamics of glaciers under their own weight in two dimensions. This seems perfect for modeling a large ice wall deforming under its own pressure in two dimensions. Halfar derived equations that can determine how the center height and the height at certain positions along the sheet would change over time at a certain temperature. The center height changes over time based on the width of the ice, the temperature, and a scaling factor that's roughly equal to 3. The second equation looks at the height of the wall as you walk from one side to the other from any starting point. If Jon Snow were to walk through the tunnel cut into the Wall at Castle Black, this equation would tell him how high the ice is above him at any point in his journey. This is based on the center height, Jon's position, the same scaling factor roughly equal to 3, and the ultimate distance he'd have to walk to get through the wall. By combining these two equations, Truffer was able to graph the Wall's center height over time, as well as what the Wall would look like if you took a cross section of it. He found that, for all temperatures, the most significant amount of deformation occurred in the first year. After that point, it slowed down over time. The rate of slowing was highly dependent on the temperature, but Truffer found that at 0°C, after that initial year's deformation, the Wall would remain roughly the same shape for the next 1,000. That part of the story is pretty accurate; indeed, the Wall has looked the same for more than 1,000 years. Unfortunately, it wouldn't look pretty. At –40°C, the Wall doesn't stabilize over time as it does at 0°C; instead, it remains taller and thinner for a longer period of time, but it continues to creep downward. At this temperature, the first 150 feet of the slope at the base of the Wall would be steep enough to

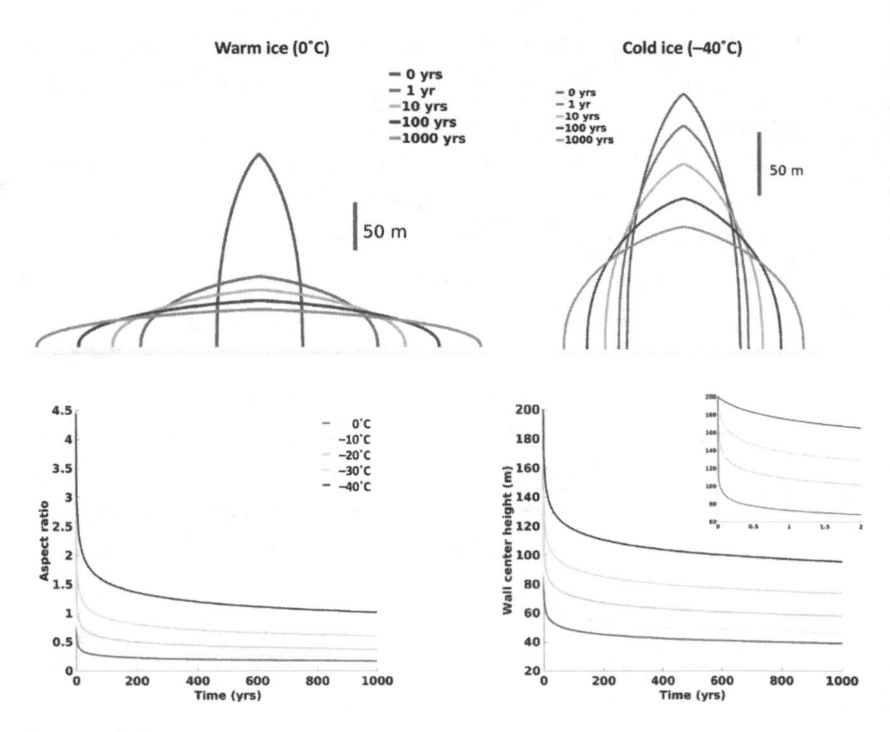

FIGURE 2.3
The shape of the Wall over time at 0°C (a) and –40°C (b), the ratio of height to width over time (c), and the center height of the Wall over time (d).
Source: Dr. Martin Truffer

prevent people from climbing it, so it might provide some protection, but the Wall would only be about 300 feet high, which is less than half the height it started at (figure 2.3).

THE GREAT WALL OF WESTEROS VERSUS THE WESTERNPORT WALL

As Dr. Truffer showed in his plots of ice wall shape over time (figure 2.3), ice just won't cut it. The wall may have been OK for a year, but over thousands of years it would be more of an ice dome or plateau, depending on the temperature. Assuming the temperature of White Walkers is around –40°C, cold enough for steel to crack easily under their grip (more on that in chapter 5, I promise), the Wall would have degraded into an easily scaled dome shape. Heck, the grade would be similar to a famous hill in western Maryland: the Westernport Wall. Every year, approximately 200 triathletes ride up the hill as part of the SavageMan triathlon. If a bunch of people in spandex can do it after a 1.2-mile swim and 19 miles into a 55.7-mile cycling course, I'm guessing wildlings and Walkers could stroll right over it, no zombie dragon needed.

But even though Dr. Truffer said the Wall likely wouldn't stay standing if it were just ice, we've got pykrete. GRRM just happened to build it into his story: the ice used to build the Wall has rocks and sand mixed into it, creating a form of pykrete—basically, reinforced ice. If pykrete can hold up airplanes, clearly it can hold up White Walkers. Unfortunately, however, there is a lot left out of this picture. First, all the amazing properties of pykrete work only at temperatures of −15°C or below. The designs for the bergship show that the Allied army intended to install a cooling system to keep the ship's temperature below −15°C. I don't think Westeros had yet developed heating and cooling systems this advanced. Pykrete (or "rockrete," in this case) may have the strength to hold up a wall, but it couldn't do it in the summer. I don't know exactly how cold it gets in the North in winter, but the average winter temperature in Alaska is between −15°C and −1.1°C. That's cutting it a little too close to the upper bound of pykrete's strength. And remember, if the wall is indeed ice mixed with rock not wood, it's going to even be a little weaker. There's a second, and bigger, problem: Pykrete doesn't stop creep. Like, not at all. It just stops crack propagation and fracture. In fact, some of the first controlled creep experiments were done during the research for the bergship.

When scientists were first testing the strength of pykrete, they rigged up a machine that would repeatedly hit a bar of pykrete with a piston. They kept the piston's speed constant, initially, but they noticed that the behavior of pykrete as it was squished depended on how quickly it was pressed. This is a hallmark of a non-Newtonian fluid. To better understand this phenomenon, they looked at how pykrete behaved over time under constant pressure by hanging a weight from a lever compressing a column of 4% pykrete with a cross-sectional area of 1 cm^2. The compression of the pykrete was measured over a number of days. The scientists found that smaller loads (approximately 7 kg/cm^2) initially experienced a high rate of creep, but it settled down to about 1% per year after that. The load of 7 kg/cm^2 was chosen under the assumption that it would be the maximum pressure the bergship would have to withstand.

The creep of the pykrete was dependent not only on the temperature at which it was held but also on the type of wood that was mixed into the ice. Canadian spruce behaved much better under load than Scotch pine. It would creep initially, but its height would eventually stabilize, just like normal ice. That doesn't sound terrible. With the addition of Canadian spruce, then, an ice wall should settle down and stop creeping—but, of course, it's not that simple. The group found that with loads greater than 14 kg/cm^2 the deformation was so bad there was no use in continuing the experiments. According to World War II engineers' calculations, a bergship with a height of about 150 feet (measured from the bottom of the hull to the top of the gunwale) can exert up to 7 kg/cm^2 on the base of the hull. The bottom of Westeros's Wall—still expected to measure over 600 feet high at its center after 1,000 years, according to Truffer—would experience about four times that pressure. This means that the bottom of the ice wall would be under a pressure of roughly 28 kg/cm^2—well past the force needed to significantly deform pykrete. It would therefore behave very similarly to Dr. Truffer's predictions.

There are no official equations for pykrete because interest in its use waned after WWII; however, Truffer's calculations are still pretty applicable for an ice wall made with stones. As Dr. Truffer said during our discussion, glaciers contain plenty of stones and other debris. Unfortunately, however, pykrete isn't the answer to Bran the Builder's problems. In the 1940s, when pykrete was being heavily researched, little research had been conducted on the creep rate of glaciers—but those berg ships were only meant to last for eight months, not 1,000 years.

It is said that Bran the Builder wove spells into the Wall during its construction. I know from *Harry Potter* that every spell has a specific purpose and a fun Latin-sounding name to go with it. Bran the Builder probably cast *Creepestis Depulso*. But even if it might not have made the best wall, ice is still pretty cool. (That joke is too easy not to make twice.)

3

North of the Wall

How to Survive in the Cold

So cold, *he thought, remembering the warm halls of Winterfell, where the hot waters ran through the walls like blood through a man's body. There was scant warmth to be found in Castle Black; the walls were cold here, and the people colder.*

—Jon Snow, *A Game of Thrones*

I love Jon Snow's hair. Really, with the exception of Kit Harington, who *doesn't* love Jon Snow's hair? His long, flowing raven locks go rather well with his perpetually brooding expression. I've never wondered why the producers keep his hair wild and free and blowing in the wind north of the Wall. I have, however, wondered how he is able to survive without a hat. According to Martin, the climate north of the Wall is pretty cold. You have also probably heard about how the *Game of Thrones* costume designers used IKEA rugs to make the cloaks for the brothers of the Night's Watch. Would a faux fur rug be enough to keep Jon warm? (Assuming Ygritte hadn't shown up, of course.) Similarly, how do animals stay warm so far north? The Night's Watch is also a fairly athletic bunch. I'm guessing animal fur (or carpeting, for that matter) doesn't wick away sweat like modern fabrics do, so would they be able to effectively regulate their body temperature while running from White Walkers? Nowadays humans can survive anywhere on Earth, from the sub-freezing temperatures at the poles to the relentless heat at the equator. But could people really survive those extremes in the days of Westeros?

The season 7 episode "Beyond the Wall" left many, many unanswered questions: Exactly how fast can Gendry run? How fast can ravens fly?

But the relevant one here is whether Jon and crew could have survived outdoors for several days in the cold, waiting for help on a small island in the middle of a frozen lake. If he was able to survive that, how did he also survive falling into frigid water and riding home in wet clothes afterward? Hopefully this chapter can answer all these questions and maybe a few more. If you've read my bio, you know that in my spare time I'm an endurance athlete. I've competed in ultra races held in the Texas and Florida heat and in Lake Placid's freezing temperatures. I've always wondered if my body could hold up to the cold in addition to the athletic feat of sprinting from the undead. I guess science will tell me if I'd be able to do it in an IKEA Tejn.

Body Temperature Regulation

Before we can understand what needs to happen to keep someone warm in the cold, it's important to understand how the human body regulates temperature and how it gets to the point where it can't. Your body has a fairly sophisticated system that keeps it at 98.6°F—plus or minus a few degrees—throughout the day, in all conditions, for your entire life. Many biology textbooks suggest thinking of it much like your home thermostat, but that wouldn't be doing it justice (although it is correct to think of it as a feedback control system). When you are too hot, systems spring into action to cool you down, and the opposite happens when you are too cold. The whole process is called *thermoregulation*. As with any feedback system, there are three parts: a control center, sensors, and a method of triggering physical and chemical changes. The control center for thermoregulation is the hypothalamus, a region in the lower front part of your brain, right above your pituitary gland. The sensors are all over your body, including in the hypothalamus. You probably know several of the ways your body mitigates heat and cold: sweating, goosebumps, shivering and teeth chattering, and digestive issues. I'll talk more about how this works in both warm- and cold-blooded animals in chapter 7, but here, I'm going to stick specifically to humans.

Here's how your body reacts when it gets overheated, either because you were outside for too long in Arizona or because you thought running (without being chased, anyway) was a good idea: First, your internal temperature starts to rise above 98.6°F. Your body can handle a small range of temperatures before it takes action, but beyond that threshold, the temperature sensors in your body alert your hypothalamus to the problem. Your brain tells your skin to start sweating to try and lose some body heat via evaporative cooling. You can get a good sense of how well this works by blowing on wet skin. The next step is one you might not notice. The brain cues your blood vessels to dilate, or expand. This allows more blood to flow under the skin and transfer heat to the skin, which is now cooling via evaporation. If you are someone who often works out in the heat, you know you swell up when you are hot for a long period of time. This is your body expanding your blood vessels. Since your digestive system is also producing quite a bit of heat, your body slows down the process and sends some of the blood from that system into the blood vessels close to the surface of your skin. This is why some people get nauseated if they work out in the heat for too long. Generally, sweating and shifting blood flow are enough to bring everything back to normal. But if you decide to keep pushing it, your core temperature will rise high enough to incapacitate the hypothalamus. With nothing to regulate it, your core temperature will continue to rise, and you probably won't be alive much longer. If your core temperature rises to about 107°F, brain damage occurs, and if it reaches 111°F—hot enough to change the structure of the proteins in your body—you'll die.

When your body starts feeling cold, your brain tells your organ systems to do everything they can to trap heat. First, your hypothalamus tells your skin to stop sweating ASAP. Then, it tells your skin to make the hair stand up to trap warm air near the body. Next, it tells your blood vessels to constrict, thus diverting more blood toward your core, which keeps it (and the rest of your body) warm. This is why your fingers seem to shrink in cold weather, and why your fingernails and lips look blue—there is less blood flow in those areas. The thyroid gland then releases hormones to trigger muscle movement to increase heat production, which is what causes you to shiver and your teeth to chatter. Your metabolism speeds up because the process by which your body uses energy is an exothermic (or heat-releasing) reaction. Faster metabolism means more heat. If none of this works and your body temperature keeps falling to 95°F or lower, you officially have hypothermia. Since the cold slows down all of the chemical reactions in your body, you first start getting confused and clumsy. The real problems start to set in when your core temperature drops below 86°F. Major organs begin to fail at temperatures that low, and death follows shortly thereafter.

It's interesting to note that, on average, your body can handle roughly the same range of temperature change in either direction. This is just an average, however, based on known statistics from real people. So, of course, there are remarkable exceptions. There are a number of people who have been revived from very low body temperatures with little long-term damage, but very few people are able to recover from severe heatstroke. (For a discussion of one of the most extreme cases of recovery from hypothermia, check out the section on drowning in chapter 13.)

This is a quick and simple explanation of how your body regulates temperature, but not really a discussion of why it needs to. That may seem obvious, but to really understand how to keep warm and how much skin can be safely exposed in extreme temperatures, it's important to look at the physical processes by which a material, including your body, gains or loses heat. There are four different ways heat is transferred: evaporation, conduction, convection, and radiation. The first line of defense against a body overheating is sweating. Sweat, which is mostly water with some salt mixed in, cools your skin by evaporating. This is why you taste like a potato chip after a run outside on a hot day. Evaporation is one of two ways liquid water can turn into water vapor (boiling is the other). The temperature of something, whether it's your body or your sweat, is a measure of how quickly the molecules are moving. The faster they are moving, the higher the temperature. The fastest sweat molecules move quickly enough to fly off and become a gas. They take energy (or heat) with them, so there is less energy near your skin and the average motion of the remaining molecules is lower. Your hot skin keeps heating up the sweat and the sweat molecules continue to carry away the heat energy on the surface of your skin as they evaporate. The surrounding air can only accommodate so much water at a time. When it's humid out, the air is saturated with moisture, so there's not much room left for the water vapor from your sweat. Since the sweat has nowhere to go, it can't evaporate and cool you off. Even though I'm an endurance athlete, I hate sweat.

As soon as I start sweating—in particular during an indoor workout with a towel handy—I wipe it off. This is a terrible idea because it means my sweat isn't evaporating. My hatred of sweat is actually short-circuiting my body's cooling method, keeping me hotter during a workout and causing me to sweat more.

The body also loses heat via conduction and convection. When a hot body is in a cooler space, the molecules on the skin are moving faster than those in the air. At the interface between your skin and the air, the faster moving molecules of your skin knock into the slower molecules of the air and transfer energy, making the air molecules move faster. Heat is flowing away from your skin to the air via molecules bumping around. This process is called conduction. Convection, by contrast, happens when fluids such as air—yes, both gases and liquids are fluids—move and carry heat with them. If you've ever heated up a frozen pizza or baked brownies, you know that a fan goes on when you turn on your convection oven. This is forced convection, meaning the fan forces air to move so the heat is evenly distributed. On human skin there is natural convective cooling. Hot air rises. When you are very hot, hotter than the surrounding air, the air closest to your skin is hotter than the air farther away from your skin. As the air near your skin heats up, it starts to rise and cooler air moves in closer to your skin. You've probably been able to see something like this when hot air rises from the hood of your car or heat rises off a hot road. In general, convection and conduction will continue until the two things in contact are the same temperature. This is called *thermal equilibrium*. A human body is continuously producing and losing heat, so, assuming the air temperature is colder than human body temperature, thermal equilibrium can't be reached while the person is alive.

The fourth way a body loses heat is through radiation, in the form of electromagnetic waves. That's a fancy way of saying you are emitting infrared radiation. You can see this type of radiation when you look at yourself through an infrared camera. Radiation is also lost via moving molecules, but by producing a type of light that is just outside the visible range. In this case, instead of the fast-moving molecules bumping into other molecules and making them move faster, their motion actually produces light. The "color" of radiation (that is, its wavelength) is based on the temperature of its source, so if you were much hotter your eye would be able to see the radiation. (That wouldn't happen until you were on fire, though, so be glad that you can only see infrared radiation via a special camera.)

How much heat is lost through each method depends largely on your body temperature and how wet you are. Radiation accounts for about 70% of the heat your body loses at air temperatures of roughly 71°F–80°F, but almost no radiation is emitted when body and air temperatures are the same. Conduction and convection account for about 15% of heat loss until the temperature reaches about 90°F, at which point sweating becomes the primary method of heat loss.[1] All four methods of cooling depend on how much skin is in contact with a colder surface. Energy (heat) will always flow toward colder areas regardless of the physical mechanism, so if hot skin is exposed to cool air, it will lose heat. The more skin that is in contact with the air, the more heat the body loses. If you are submerged in water, heat loss happens much more rapidly. As I mentioned in the previous chapter, water has a very high heat capacity and conductivity. This means it can take a whole lot of heat without changing its temperature very much, and it does this very quickly. Your body will continue to lose heat until thermal equilibrium is reached, so it is much harder to heat the water close to your body to the same temperature as your skin. Because of this, your body will lose heat about 20–30 times quicker in water than in air. If you are trying to avoid being cold, try to stay dry. Indeed, the last and best advice to any polar explorer is: "Stay dry—if you get wet, you're dead."

All of the metabolic processes going on inside your body produce heat—much of it in your core—and the more volume there is to be metabolized, the more heat will be produced. That heat has to be dissipated if your body needs to be cooled down. Heat loss is determined by how much skin is exposed to the air, which depends on a person's clothing choices and their body's surface area. The higher the ratio of surface area to body volume, the more quickly a body can cool down. As something increases in size, be it a ball or a polar bear, the surface area increases much more slowly than the volume does, thus decreasing the surface area–to–volume ratio. A tennis ball has a surface area–to–volume ratio roughly 10 times that of a yoga ball. It is therefore much harder for a large animal or person to decrease their core temperature because they just don't have enough surface area to get rid of all the extra heat they are producing. This is not so great during an Arizona marathon but fabulous north of the Wall. The phenomenon of animal size as it relates to climate is summed up in Bergmann's rule, which says that the average size of a species depends on the climate, with larger animals living in colder climates. This is also why pregnant people get very hot. As their volume increases, their surface area doesn't increase quickly enough to get rid of the heat produced by the cute little metabolic furnaces growing in their belly. My ex's parents are both scientists, and when his mother was pregnant with him, she would sleep with the air conditioner on full blast while her husband shivered under several layers of blankets. He found this ridiculous and decided to use science to "prove" that it was all in her head. He took measurements of her growing belly and calculated her surface area–to–volume ratio. He concluded that she was, indeed, as hot as she felt. (How she concluded not to kill him, I'll never know.)

When Your Body Just Can't Take It

If everything is going according to plan, the temperature regulation systems in your body keep you very close to 98.6°F. That isn't always the case, however. Sometimes you end up in a place so hot or cold that your body can't effectively dissipate or retain the heat. Because this chapter isn't about Dorne, I'll stick to talking about what happens when your body isn't able to keep itself warm rather than how it reacts to overheating.

According to the US Centers for Disease Control (CDC), roughly 1,300 people die from excessive natural cold each year. (I'm guessing "unnatural cold" means being stuck in a freezer.) About twice as many men as women die of hypothermia, and the number has been on the rise as of late. As you would expect, death by hypothermia is more common in children and the elderly. The average temperature when people died was 6°F. Though the rate of hypothermia deaths (per 100,000 people) has almost doubled, it's still a pretty rare way to go, partially because a patient has a pretty good chance of recovering as long as they are found and treated in time. There are number of other causes of hypothermia, but I'll limit this discussion to death from exposure.

Hypothermia can be mild, moderate, or severe depending on the core body temperature (90°F–95°F, 82.4°F–89.9°F, and <82.4°F, respectively). As I've already mentioned, when your body first starts to sense its temperature is dropping it tries to combat the drop by kicking your heat-producing systems into overdrive. Your blood vessels constrict, and your skin gets pale. Your body releases thyroxine and epinephrine to increase your metabolism and your heart rate, inducing other systems to produce heat. You'll shiver, want to run around, and probably feel hungry and like your heart is racing. Shivering alone can cause a two- to fivefold increase in the rate of heat production. You will also have to urinate more because increased metabolism means increased waste production.

Your body only has so much fuel to sustain this energy production, however, so at some point these systems start to slow down. Your brain will become sluggish and you'll start feeling apathetic and confused. You'll lose coordination as your muscles become exhausted and blood flow to your brain is reduced, seriously impairing your judgment. You probably shouldn't be making any important decisions while this is happening. (In other words, it's probably best not to say yes to a marriage proposal or agree to a military alliance if you are hypothermic.)

If you don't start to warm up soon, your body will move into moderate hypothermia, which causes all of your organ systems to slow down. Your body doesn't have enough fuel, so it slows down to preserve what's left. You stop shivering. Your heart rate decreases. Your breathing slows. Now your brain is no longer getting the oxygen it needs, so you are less conscious of your surroundings and your coordination is markedly impaired. You may begin to hallucinate. Blood-clotting proteins don't work if your body temperature is too low, so if you hurt yourself, you may not stop bleeding.

If you haven't yet found warmth, you will move on to severe hypothermia. At this point, you will likely slip into a coma. You will only breathe intermittently, if you continue breathing at all. Your brain will cease to function, as will your pupils. You'll no longer urinate. Your heart will begin to beat erratically and then stop altogether. Right before death occurs, your blood vessels will no longer be able to stay constricted due to lack of energy and will begin to dilate. Blood will rush to your skin and you will feel extremely hot. Many people, if they are even mildly conscious, begin to undress at this point, a phenomenon called *paradoxical undressing*. At this point one in four people will engage in *terminal burrowing*, which is exactly what it sounds like. They will either create or find a tight space and crawl in there to die. The cause of this behavior is unknown. As far as causes of death go, even though it is drawn out, this one is described as not too bad. I had hypothermia once after a long, cold swim. Of all the symptoms, the worst part for me was the confusion and the lack of coordination. Getting a wetsuit off while not being able to comprehend how the zipper functions or even stand on one leg led to many blackmail-worthy photos. Many people have been found quickly enough to survive severe hypothermia and can recount the experience. In one case, Brian Phillips, a reporter following the Iditarod, was stranded in the Alaskan tundra for a few hours when his plane's engine froze. He described the experience in a piece for ESPN: "It didn't feel violent, that was the thing. Even with the wind ripping past you. It was like certain parts of your body just accrued this strange hush. Like you were disappearing piece by piece."[2] After researching and writing chapter 13, I'd have to say that, of the myriad ways to die, hypothermia would be one of my top ways to go.

Luckily for Jon and crew, as well as countless others who have been in cold situations, recovery from severe hypothermia is common. In general, the treatment for hypothermia is to get somewhere warm and let your body heat up again. You may have seen athletes at the end of marathons or Ironmans wrap up in space blankets. These athletes usually get cold very quickly because they are no longer producing extra heat through activity, but their bodies haven't yet turned off the cooling mechanism. Space blankets reflect the body's radiant heat and trap an insulating layer of warm air around the body. This is the first line of defense in potential or mild hypothermia cases. Moderate or severe cases require a trip to the hospital. Forcing a hypothermia patient to warm up through lots of intervention is often counterproductive. Moving them to a warm location and letting their body do its thing is often the best course of action. In severe cases, warming blankets and heated IV fluids are used. (The fancy way to heat an IV is to stick it in the microwave, in case you were wondering.) In extreme cases, doctors may insert tubes into the front and side of the patient's chest and flush it with heated saline to aid in warming the internal organs. With hypothermia, even patients who appear to have died can still be revived. Sometimes rewarming is prematurely discontinued on a patient because they don't appear to be responding, but in reality, they are still alive and stopping treatment is actually what causes death. As the saying goes, "Nobody is dead until warm and dead." Assuming someone with hypothermia is warmed to a safe temperature before they are beyond resuscitation, they have a good chance of survival.

ANIMAL FUR (EVOLUTION IS AMAZING)

Though surviving hypothermia is common, the best course of action is to, well, not get hypothermia in the first place. For something to effectively insulate the body, it needs a way to stop all the ways the body cools off: sweating, convection, conduction, and radiation. The easiest adaptation is to stop sweating. The rest need some sort of external help, whether that be a fur coat, a layer of fat, or an IKEA rug. Humans have evolved to survive in a range of temperatures, but polar bears, moose, yaks, and others have evolved to thrive in it. As previously mentioned, they are large, which means a low surface area–to–volume ratio. They lose heat more slowly relative to the heat produced by body functions than smaller animals. These animals also have dark-colored skin to be able to absorb and retain heat as much as possible. They have a layer of fat to better keep in heat and keep out cold. These animals all make use of a two-coat system. Their undercoat is coarse and dense and designed to insulate in both high and low temperatures. The top layer is composed of guard hairs, which are longer. In this layer, color is the most important feature, but it can also serve as a sensor, alerting the animal when something is close enough to brush a hair but not a body, and add to thermal insulation. Since staying dry is an important factor in keeping warm, the top layer is often coated with oil to repel water. In short, they are pretty well equipped to conserve as much heat as possible.

It's worth talking about the science of two of these adaptations, fat and fur, in more detail. There is a general intuitive understanding that fat insulates. Whether that (correct) assumption comes from seeing large animals in cold climates or knowing that larger people are generally hotter, I don't know, but the physics of why the assumption is correct is rarely discussed. There are a few ways fat aids in warmth. Two of these help the body produce more heat and one counteracts the effects of conductive heat loss. First off, fat has a very low heat conductivity. It transfers heat four times less effectively than muscle tissue. This means that heat can't get from the body's core to the air or water outside.[3] Second, fat acts as a way to store energy. A body requires a large amount of energy to sustain the processes needed to produce enough heat, so an animal needs reserves of stored energy. Third, fat does something really surprising that was just recently discovered: fat can sense cold and respond.[4] When fat gets cold, it produces a protein called UCP1 (uncoupling protein 1). This protein in turn uses fat to produce heat. This was an incredible discovery because it led to a better understanding of how fat keeps animals warm and also because, as its name suggests, it demonstrated how it may be possible to break down fat. A better understanding of how fat cells sense cold and produce UCP1 may lead to improved treatments for diseases such as type 2 diabetes and obesity. The effects of fat on insulation are particularly important with regards to water, given that bodies lose heat much more quickly when submerged. Aquatic mammals such as seals and manatees have very thick layers of fat. Long-distance swimmers are known for actively gaining weight to help insulate their core in extreme conditions. When swimming across large bodies of water, such as the English Channel, there is a huge risk of hypothermia both because of the chilly water and because the rules of official open-water swims prohibit wetsuits or any insulation beyond a swimsuit. This is no small feat when you consider that training involves swimming many miles each day and burning lots of calories. An open-water swimmer's diet often consists of pints of ice cream. This is normally where someone would make a joke about open-water swimming sounding like a great sport. Having done the swim from Alcatraz, however, I am confident that no amount of ice cream is worth trying to survive for 13 hours in frigid water.

The structure and function of fur is crucial to keeping an animal warm and is an amazing feat of evolution. Animals' coats are designed to counteract heat loss from water, conduction, convection, and most importantly radiation. Fat is only able to mitigate the effects of two of those. An animal's dual-layer coat prevents heat loss by conduction and convection by keeping a layer of air close to the animal's body. The undercoat is so dense that air gets trapped between the hairs and is warmed up. Because the air can't circulate, no heat is lost to convection. The trapped air insulates the animal's body and prevents heat loss from conduction as well. In addition, many northern animals, including polar bears and moose, have fur composed of hollow hairs. It was initially thought that this would channel the heat of the sun to their skin much like a fiber optic cable, but this theory has been thoroughly debunked. Hollow hair is actually yet another barrier preventing heat loss. Many of the animals that live in extreme cold are also partially aquatic, so it is crucial that their coats be able to provide the same protection in the water. This is done in several ways: the coats are dense enough that water can't penetrate all the way to the skin, small glands next to each hair follicle coat the hair with oil to make it repel water, and the hairs are shaped in such a way that they lay flat when wet to better trap air as the animal moves in the water.[5]

These properties of fur effectively protect against heat loss by conduction and convection, but not radiation. Since most body heat is lost by radiation, a type of fur that could effectively stop heat loss via radiation would be a huge evolutionary advantage. Until recently, scientists thought that hair only blocked heat loss by convection and conduction, but computer simulations by Dr. Priscilla Simonis, a physicist at University of Namur in Belgium, showed that thermal radiation is also contained. Though polar bears look white, their hair is actually translucent. Polar bears' hair color (or lack thereof) serves two purposes: first, it helps them blend in with their surroundings giving them excellent camouflage; and second, because it is translucent, it reflects all wavelengths of light—including infrared. When radiation is emitted from a polar bear's skin, each hair reflects it in a random direction. Because there is so much hair to scatter the radiation, the majority of it is trapped in the coat, thereby keeping it warm. Their fur traps radiation so effectively that the bears can't even be seen with an infrared camera—they are *that* insulated. This makes polar bears really difficult to track via plane. They blend in with their surroundings and can't even be detected by a thermal camera.

Animal adaptations to the cold are pretty amazing, but could humans—and, in particular, those living in the age of Jon Snow—create something as effective? Or would they be better off simply improving their hunting skills?[6]

Keeping Humans Warm

Equipping yourself for cold weather, especially if you're planning to be active, is a balancing act. Clothing needs to be warm, but not too warm—sweating should be avoided at all costs. It must be tight enough to stop air from getting in or out, but not so tight that there's no room for a layer of insulating air. It needs to repel water without trapping the moisture produced from your body. The key, just as your mom always suggested, is to dress in layers. Usually, this means a base layer that wicks away moisture, a thick middle layer that acts as a barrier between your body and the outside to prevent heat loss by conduction and convection, and a waterproof outer layer. There are many modern techniques used to stay warm, and it's worth looking at a few of them because the science is pretty interesting. The Rangers of the Night's Watch are a pretty athletic bunch of guys, so I'm going to talk specifically about clothing designed for being active in the cold. The Night's Watch probably didn't have access to Gore-Tex, however, so it's more useful to talk about what they do wear: wool, leather, quilted fabrics, and animal fur.

Just like with animal fur, clothing for outdoor activity needs to stop or slow down all four ways the body loses heat. Stopping heat loss from convection and conduction is usually done the same way as with animal fur, by creating a layer of still air close to the body. When you are working hard outside in the cold, your body will still sweat. Any high-tech cold weather gear needs to be able to wick away that sweat without cooling you down, for two reasons: first, it's uncomfortable and can lead to chafing if that moisture is held close to the skin; and second, you will cool down very quickly as soon as you stop moving around, potentially putting you in danger of developing hypothermia. We know that getting wet is the quickest path to hypothermia, so as the material pulls moisture away from your skin, it also needs to keep out moisture, such as rain and snow. Gore-Tex, invented in 1969, pioneered this type of wicking yet waterproof clothing. Named after its inventors, Wilbert and Robert Gore, it was the first high-tech fabric to claim it could repel external liquid water and at the same time allow water vapor from sweat to pass through its fibers. Gore-Tex is a novel use of Teflon, which was invented in 1938. The thread used to weave the fabric is made by heating and then quickly stretching the Teflon to about 800% of its original length. This creates spaces between the microstructure of the Teflon. The stretched material is about 70% air, which helps with insulation and reduces weight. The microstructure of stretched Teflon, not the molecular interactions between water and Teflon (figure 3.1), is what gives Gore-Tex its wicking

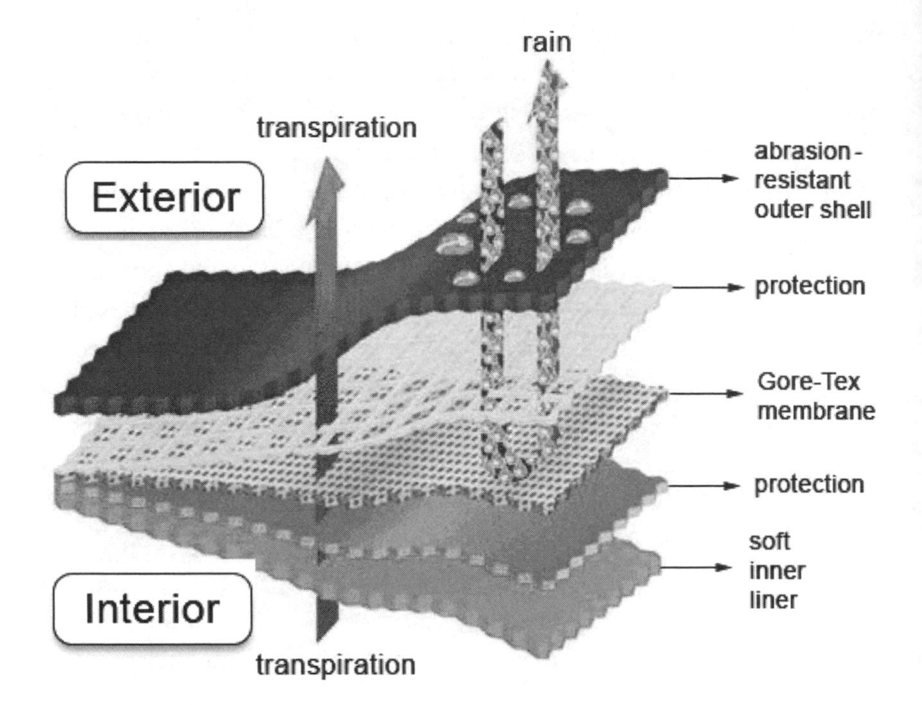

FIGURE 3.1
Diagram of Gore-Tex. Notice that it's the Gore-Tex that is waterproof, not the exterior of the fabric.

yet waterproof properties. Water vapor molecules are much smaller than water droplets. Sweat evaporates and turns into water vapor, which can escape through the Gore-Tex. Water droplets, however, cannot make it through the fabric. If the Gore-Tex ends up completely drenched with water, however, it loses its wicking ability because the water prevents the water vapor from escaping. Because of this, Gore-Tex clothing is made with several layers: a soft inner layer, a layer of Gore-Tex, and a water-resistant layer that will still allow water vapor to escape but isn't as water resistant as Gore-Tex. The clothing is waterproof because of the Gore-Tex core, not the shell, but the shell keeps enough water away from the middle layer to keep it breathable. Gore-Tex was the first material able to strike the right balance between warmth and wicking, but it did not address the issues of radiative heat loss.[7]

Since radiative heat is the primary way that heat escapes the body, it would make sense to stop heat loss from radiation. To date, Under Armour seems to be the only athletic clothing company to produce a product specifically designed to trap radiated heat: the aptly-named Infrared. The fabric is treated with a thin coating of ceramic to trap and reflect radiated heat, much as a coffee mug does. This technology made quite the splash when it debuted, with advertisements touting the fabric's ability to absorb and retain body heat. That claim has not been substantiated by users, but it has amassed quite a following nonetheless. I personally own a pair of Infrared running tights as well as a compression shirt, and they are my go-to cold weather running gear. Unlike most other high-tech fabrics, which are used only in certain layers of clothing, Under Armour applies Infrared technology to everything from the base layer outward—even accessories like gloves. I don't want to sound like a shill for Under Armour, but I do want to specifically highlight the fact that they are currently the only company attempting to combat the number one method of heat loss. The limiting factor is not how to stop this kind of heat loss but how to do it while simultaneously wicking away sweat, allowing the wearer to move freely, insulating them from the cold, *and* reflecting body heat emitted via radiation. The most ubiquitous safeguard against radiative heat loss is the space blanket, but Mylar isn't exactly comfortable athletic wear.

There is an even newer technology that has recently hit the market with the goal of responsive insulation. Both Under Armour, through their new Reactor technology, and the North Face, with their Ventrix line, are in the game this time, but using different technology. Both use "adaptive materials," meaning their properties change based on external stimuli. In the case of Ventrix, this stimulus is stretch. When you are exercising, you are producing heat and thus need more ventilation. The Ventrix material has vents that open up as you stretch and move around, helping to cool you down when you are active. When unstretched, however, the vents close and provide the insulation you would expect from a thick jacket. In this way, it can not only stop you from overheating when exercising but also keep you warm once you're done.[8] Under Armour approached the challenge in a different way, or at least I think they did; they are extremely tight-lipped about their tech, and reasonably so. What

I have learned, however, is that it uses a patented insulating material that changes its absorbency relative to the amount of sweat your body is producing. In addition, the layers of fabric are stitched together in a way that places more insulation around parts of the body that are most likely to get cold during outdoor exercise and less insulation around the body parts that are likely to get hotter. Together, the insulation and stitching pattern promote air flow and offer improved wicking properties. Athletic wear companies are continuing to develop new and improved methods of striking that delicate balance between being warm and being comfortable while exercising. So, if you are using winter weather as an excuse not to run, you might not be able to keep that up for much longer.

All of these high-tech materials are fascinating to me as a scientist, but obviously the Night's Watch had to rely on outerwear made from materials such as wool, leather, animal pelts, and woven fabrics. Nature is really quite amazing, however, so these low-tech options were still pretty good. Of these options, leather isn't great at keeping heat in, but it's really great at keeping out things that might cool someone off. Leather is particularly useful for keeping out moisture and wind, both of which have a significant effect on heat loss from convection and evaporation. Leather gets its unique properties from the structure of animal skin as well as from the tanning process. For something to be both wind- and water-resistant, it needs to have fibers that are bound tightly together. Animal skin is not uniform in its structure; the skin around joints is more flexible because its fibers aren't bound so tightly, whereas the skin covering areas such as the upper back is stiffer and has a tighter fiber structure. The highest-quality cow leather comes from the animal's back, along the spine; by contrast, the shoulder joints provide the lowest-quality leather. The tanning process—which involves conditioning the leather with oils, waxes, greases, and vegetable products to fill in any gaps between the skin fibers in an animal hide—affects many of the properties of leather, including breathability and water resistance. The weather-resistant properties of the final product vary depending on the tanning agent and the quantity used. In general, however, leather is known more for keeping out wind and rain and not so much for its general warmth. It also has the added benefit of being difficult to cut through with a sword.

Thick fabrics are better for keeping heat close to the body, though this wasn't initially their purpose. The *gambeson*, a padded jacket constructed from linen or wool and stuffed with anything from horse hair to scrap cloth, was worn either on its own or under mail or plate armor. It was not designed for warmth; in fact, the heat it trapped was a drawback. They were typically worn to prevent chafing from metal armor, but they provided a good defense on their own—thicker varieties could even prevent heavy arrows from piercing the wearer. The Night's Watch can often be seen wearing them, but unlike others, they would have benefited from the gambeson's warmth in addition to its defensive quality. Linen is a naturally wicking fiber, so if it were used to make the gambeson, it would draw sweat away from the skin, thus reducing heat loss by evaporation. The horsehair stuffing would provide an extra layer of thermal insulation much like polar bear hair does, creating still air pockets that trap heat as the body warms up and serve as insulation. Horsehair is also incredibly durable, adding to the protection offered by the gambeson. These garments do double duty for the Night's Watch: as a more manageable (both financially and in terms of weight) alternative to armor and as insulation against the elements.

Animal fur is an amazing insulator, as I mentioned earlier, so it makes sense that the Night's Watch would use this evolution-approved technology to their own advantage. Fur was one of the first materials used for clothing about 70,000 years ago, around the end of the first ice age. Animal pelts can be extremely warm for all the reasons listed here. The fur properties now keep the wearer, and not the unlucky bear, warm. The leather underneath keeps it wind-proof and water-proof. Unfortunately, it isn't quite as good on humans as it is on animals. A group led by Heather Liwanag, then at the University of California, Santa Cruz, found that the thermal conductivity of a pelt (fur and skin) was much higher than the thermal conductivity of fur alone. This means that more heat would be lost through a pelt worn by Jon Snow or Sam Tarly than by a bear. Of the animals her team analyzed, seals and walruses have pelts with the highest thermal conductivity, meaning they insulate the least. Bear pelts, which look a lot like what the Night's Watch uses, are about twice as warm as seal pelts. The warmest pelts of all, however, come from animals in the Procyonidae family, which is a fancy way of saying raccoons and kinkajous. Wolf pelts come in somewhere between the two. One of the biggest drawbacks of using fur as the main source of warmth is the weight. A grizzly bear pelt weighs roughly 24 pounds. That isn't a lot, but it's far heavier than something like wool. The Night's Watch uses fur primarily as trim on their shoulders and upper chest. This would help keep their arms warm and allow them to adjust the warmth of their core based on their activity level. For reference, the IKEA Tejn rug weighs about a pound, according to the shipping weight on IKEA's website. It's made mainly of polyester fiber, which has lower thermal conductivity than a bear pelt does, and it sells for $14.99, which I think is probably cheaper than the drawn-out hunting expedition that a bear pelt would be. The costumers did the Night's Watch a service by substituting the Tejn for a bear pelt, though I think its wicking properties might be nonexistent.

It appears that the last element of the Night's Watch uniform is made up of wool cloaks. Wool is quite possibly nature's super material. Wool is warm, wicking, flame-retardant, flexible, and light. Many of wool's superpowers come from the structure of the fiber. Wool fiber is coated with scales (figure 3.2), which are a key part of their ability to insulate and wick away moisture. If you are a knitter, you've heard about a process called felting, where a knit wool garment is wetted and agitated until the yarn locks together to create a solid, thick fabric. The water causes the scales on the outside of the fiber to stick out and the agitation causes them to lock together, forming a thick mat. The air pockets in felted wool garments trap warm, still air, thereby stopping heat loss via convection and conduction. The resulting felted wool has a heat conductivity about the same as a bear pelt, but a wool cloak would weigh only about 10 pounds, making it significantly lighter and more flexible, not to mention more sustainably and safely acquired.[9] In addition to its warmth, wool wicks water vapor like Gore-Tex but in a slightly different fashion. The outer layer of a wool fiber, the cuticle, has tiny pores that allow water vapor to move into the core of the fiber. This absorption involves a chemical reaction that produces a bit of heat, so when wool is wicking, it is also heating. Wool can absorb up to 30% of its weight in water and still feel dry, whereas synthetics can only absorb about 7% of their weight. In addition,

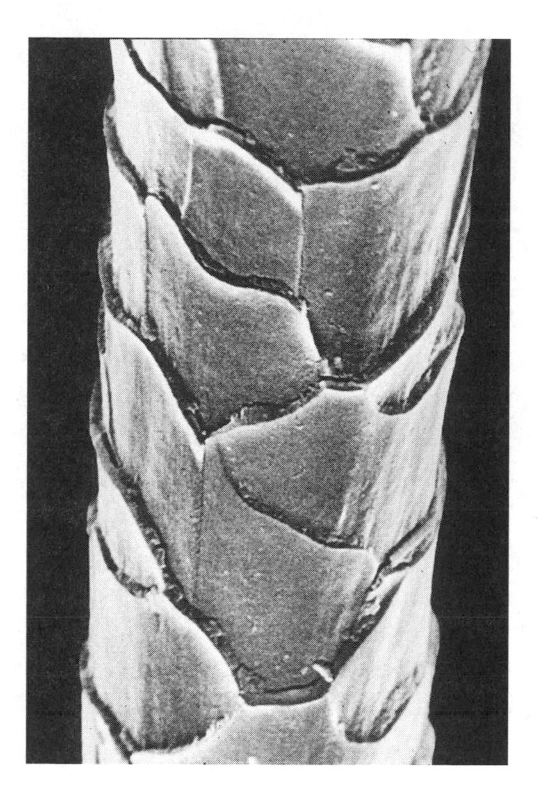

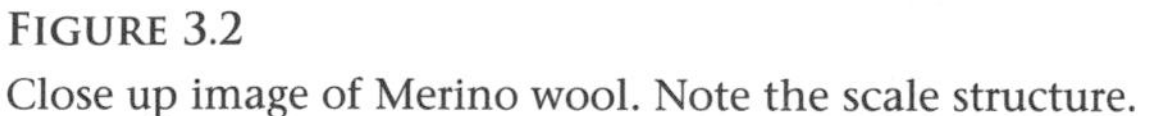

FIGURE 3.2
Close up image of Merino wool. Note the scale structure.

wool is flame-retardant, so it is often used in fire blankets. I have a side gig as a pyrotechnic engineer for a fireworks company, and they always have wool blankets on hand in case something goes wrong. Because of its high water content and chemical structure, wool isn't able to sustain a combustion reaction (more on that when I talk about dragon fire). Fire requires a significant amount of oxygen to sustain itself, so even if the wool catches fire, its water content will douse a flame before it can really get going. The main drawback of wool is that the same scale structure that makes it so useful also makes it rather itchy. In addition, about 20% of the population is allergic to wool altogether. These unlucky people will not only have a skin reaction but respiratory reactions as well. (As an avid knitter, I'm very glad I'm not part of this 20%.)

DOES JON SNOW NEED A HAT?

All of the discussion until now has been theoretical, based on several research papers. Personally, I'm a big fan of doing endurance sports in warmer climates, and living in Maryland hasn't afforded me the opportunity to train in temperatures like those in the North. Before discussing what Jon and crew would have potentially experienced on their season 7 journey north, I wanted to talk to someone who has done something similar in real life. I was lucky enough to have a conversation with the marathoner Beth Sanden, the first physically challenged athlete to complete a marathon on all seven continents plus the North Pole. She is an incomplete paraplegic and an endurance sport legend. If anyone knows how to deal with a workout in the cold, it's her. She spoke extensively about her experience in the Antarctic and North Pole marathons. Both were very different in weather and terrain. She completed the Antarctic marathon in 10°F weather in about 11 hours, and the North Pole marathon in 7.5 hours in −50°F weather. The climate of both poles was very dry, so body cooling from external moisture wasn't really an issue, and it was the terrain rather than the weather that was the limiting factor in Antarctica. Sanden spent months training for the marathons in the mountains to acclimate herself to the cold. In both events, she wore a type of tech I haven't yet talked about: battery-heated hunting gear. This worked great in Antarctica, but it was so cold at the North Pole that the batteries in her goggles malfunctioned and her eyelashes froze. Before I spoke with her, I had assumed hypothermia would be the problem because she was racing in such low temperatures. None of the competitors in either race developed hypothermia, but frostbite was a real issue. It took a year for her dead fingernails to grow back after her fingers got frostbite at the North Pole. The first thing the human body does when it gets cold is divert blood from the skin and extremities, so it is easy for the cells to freeze and die, which is what causes frostbite. The adaptation seemed to work pretty well, however, considering everyone finished with a reasonable body temperature. I would highly recommend reading her story online—she is a true inspiration and, put simply, a badass. In 2018, she competed in the Boston Marathon. For those of you who follow endurance sports, you know that was a damn tough year in Boston.

Sanden said that, without question, the 2018 Boston Marathon was more miserable than the Antarctica and North Pole marathons. The reason? The rain. As I've mentioned, water saps the heat out of your body 20–30 times faster than air due to its high heat capacity. Competitors in the Boston Marathon faced temperatures in the mid-to-upper 30s as well as freezing rain and wind—every single factor that can steal heat from your body quickly. There were more hypothermia cases at the 2018 Boston Marathon than there were in either of the extreme cold weather marathons. The weather was not cold enough to freeze flesh, but the water effectively cooled down the body to the point that many runners required medical intervention for hypothermia. Keeping all that in mind, let's look at Jon and his trip north to capture a wight.

Before I go on, I think we need to collectively acknowledge the fact that Jon, Dany, and everyone else managed to concoct one of the stupidest plans in recent fantasy history. Venturing outdoors in subfreezing temperatures, near a lake, with minimal supplies, to grab a zombie intent on killing them—all without a hat. I think we can all assume two things: one, Jon is not the quickest raven in the unkindness; and two, science will show they all would have ended up dead. The two science questions I'd like to answer as I close this chapter are: Exactly what would have killed them, and how much did their lack of hats have to do with it? I'll answer the second question first.

There is a common myth that people lose 70% (or 60% or 40%, depending on whose dad you ask) from their head. From a quick-and-dirty science perspective, it makes sense that heads would lose heat more quickly than the rest of the body. First, there is quite a lot of blood flow to the head, and because your body needs the brain to stay alive it can't shunt blood away from the head to conserve core temperature as easily as it can from your fingers. The large volume of blood flowing through the head loses a lot of heat to the air, causing the core temperature to drop when the blood circulates back through the heart. The head has very little fat to use as insulation that would stop heat from escaping by convection or conduction. Additionally, the head has a larger surface area–to–volume ratio than other parts of the human body. But do all these factors really lead to relatively more heat loss from the head? Thea Pretorius from the University of Manitoba wanted to test this myth. She submerged volunteers, some up to their neck and some fully, in 62°F water and looked at their rate of heat loss. Those with their head submerged lost only about 10% more heat than those who were allowed to keep their hair dry.[10] Pretorius dispelled the "mom myth" of heat loss caused by not wearing a hat, showing that the head loses heat at about the same rate as the rest of the body. This can also be shown by looking at an image of a human taken with a thermal camera; as you can see in figure 3.3, the head isn't radiating any more heat than the rest of the exposed skin. Nevertheless, wearing a hat is still extremely important. Interestingly, when the head is the only body part exposed to the cold, the body doesn't react by shivering. This causes your body to lose heat much more quickly in cold weather than you would if your shiver reflex was triggered. Studies have shown that the effect of the volume of blood in the head isn't insignificant. Body temperature will drop much more quickly if someone decides to bundle up but leave the hat at home. In

FIGURE 3.3
Thermal image of a human. He is losing the same amount of heat from his hands as he is from his head. Heat loss from his body is reduced by his shirt.

addition, the ears have very little blood flow and are prone to frostbite if not covered. So, the myth is right, but for unexpected reasons, and yes, Jon would be very cold without a hat. Considering he's about to spend several days on a rock in a middle of a freezing lake that he will eventually fall into, a hat would be a really good idea. For those wondering: no, his hair does not count as a hat. Hats work by keeping warm air trapped near the head. His hair blows around in the breeze, so it is not stopping a chilly draft from blowing across his scalp. I'm not happy about him having to cover up those luscious locks, but if we want him to live, a hat would be a great idea.

There has been quite a bit of fan debate about how long Jon and company were stuck on the rock in the middle of the lake. To minimize their life-threatening time on that cold rock, Gendry would certainly have to be the first to achieve a sub two-hour marathon (have you *seen* Gendry?), ravens would have to fly more quickly, and dragons would have to be actual giant supersonic demons. The best way to estimate the time elapsed is by estimating how long it would take the ice to refreeze enough to bear the weight of the wights. This would take days, not hours. In 20°F weather, ice-fishing guidelines suggest that a lake would freeze about 1 inch per day. About 5 inches of ice would be needed to support the weight of a snowmobile, so I'm going to guess that 5 inches is also thick enough to withstand a battle scene.[11] This means that our fearless idiots would have to be stuck on the rock for about five days. A human can survive for up to 40 days with little to no food as long as they are hydrated, so the group likely wouldn't have starved to death. As hypothermia set in, however, they would probably have to pee quite a bit. In addition to being weak from lack of food, let's consider their likely fighting condition after five days in the cold. First, our hatless wonders would get frostbite in their extremities within in a few hours, based on their lack of proper head covering. Here is where things get trickier. It is not easy to estimate how long it will take for someone to die of hypothermia because everybody reacts differently and the rate of heat loss depends on the clothing they're wearing. The body's natural response to cold is to start burning calories. This means our team would need more food to keep their body temperature up. Unless they were hiding something, our team didn't carry much with them into the wild. This would both speed up their body's cooling and decrease the time it would take for them to starve to death. The fact that they suffered through roughly 11 nights means they were subjected to even colder weather. Although I can't say exactly how long it would take for them to die, there is no way they would have lasted 12 days in the bitter cold clothed only in animal furs, leather, and wavy hair. Even if by some miracle they did, they would be in no state for an epic battle. The entire party would be dead or comatose by the time Daenerys arrived. Jon, however, who seems to seek out death the way I seek chocolate, wanted to try one more time to die. He was in the frozen lake for roughly a minute or two, if you assume the sequence was shot in real time. The water was at freezing temperature; we know this because of the ice. The longest someone has withstood freezing temperatures was a drowning victim who was submerged for almost an hour and ultimately survived. Victims of the sinking of the *Titanic* were reported to have died in as little as 15–30 minutes. Jon's time in the lake would not have killed him; however, he emerged soaking wet and then rode back to the Wall. I have no idea how long the journey would take, but Jon—soaking wet, in below-freezing temperatures, with either natural wind or wind created by riding on horseback—probably would have had less than an hour in his quest to get back to warmth. I'll assume no one would find him if he fell off his horse, so that means he would have to be able to stay on a horse until he reached them all. He wasn't conscious when he got back to the Wall, but he was alive. It's not clear how far beyond the Wall the group had traveled, but I know it must be more than an hour on horseback. Maybe he is one of the lucky victims of hypothermia who isn't dead until he's warm and dead. That would give him maybe two or three hours. Against all odds, and assuming they didn't have to go far to find White Walkers, Jon could have made it home alive—if he didn't get frostbite from not wearing a hat, that is.

4

White Walkers, Zombies, Parasites, and Statistics

There are darker things beyond the Wall.

—Catelyn Stark, *A Game of Thrones*

Zombies are the vampires of the late 2000s and early 2010s, but they have a much richer history, having been a fixture in entertainment for many years. As with the vampires before them, zombies have a generally agreed-upon set of rules that writers follow, but the rules can be bent to suit the story. Heck, vampires go from being soulless, deadly assassins who die in sunlight to having a soul and falling in love with slayers to not only surviving in sunlight but *sparkling*. The rules are even more flexible when there's romance involved. The same is true for zombies, with different narratives imposing different rules. With roots in voodoo, *White Zombie* (1932) is widely considered the first zombie movie and has very clear rules for how zombies behave. George A. Romero's *Night of the Living Dead* (1968) is seen as the starting point for modern zombie lore as well as the idea of the "zombie apocalypse," though the term "zombie" is never used in the film. With a few exceptions, the current boom of zombie films seems to feature creatures that are slow and lumbering, don't have a zombie master of any kind, and are generally in search of brains (to eat—no Scarecrows here). The common factor, though, is that zombies are undead, meaning animated but not alive. They are controlled by something—either a puppet master figure or an insatiable craving for brains—and have some ability to move and attack.

Sounds a lot like some of our cold friends north of the Wall.

And just like the rest of America, scientists can't seem to get enough of zombies. Of all the fictional monsters, I believe zombies are the most studied. There is research into the neurology, anthropology, statistics, mathematics, biology, and diet choices of zombies—so much research, in fact, that there is even a Zombie Research Society that collects and distributes all of this work.[1] They have a very conservative view of what a zombie is (see the definition in the next section), but if you are looking for current research on all things zombie, this is the place to start. Unfortunately, their membership link is broken or I would have joined.

Zombies aren't just fictional. Often, there will be headlines about "zombie raccoons" or "zombie wasps," with articles discussing how an actual animal can turn into a zombie and what that behavior might look like. Numerous websites and research papers have been devoted to discussions on how to survive a zombie apocalypse. In fact, there is more than one zombie apocalypse survival kit available for purchase. With all this zombie love and lore going around, it is my hope that in this chapter I can put the wights and White Walkers of *Game of Thrones* in a "scientific" context. How do they operate? How do they move and fight? How does this compare to real-life zombie animals? Wouldn't their bodies decay? Most importantly, what can current statistical research tell us about how to survive an encounter with the army of the dead if a few of us rogue adventurers were to find ourselves north of the Wall without dragonglass?

What Is a Zombie, and Do Wights and White Walkers Count?

This really does seem like it should be an easy question. Zombies are neither dead nor alive; they are generally decaying corpses; they want to eat brains; and they don't have much coordination. It's possible, however, to find an exception to every one of these rules. Disney zombies sing and dance, but this is ridiculous so I won't discuss them further. According to the Zombie Research Society, zombies are defined as "a relentlessly aggressive human, or reanimated human corpse, driven by a biological infection." They go on to state that "the zombie pandemic is inevitable, and survival of the human race is crucial. It's simply a matter of when . . . so be prepared." So, by their definition, zombies are aggressive and obey no master—they are driven only by the need to spread their infection. When it comes to zombies, it seems they can be sorted into classes based on a few key questions and attributes. The first question is: Who or what is controlling them? There are have been accounts of zombies either operating independently, as in *Night of the Living Dead,* or being controlled by a single person, as in *White Zombie.* Another question that needs to be answered is zombie motor skills. In the case of iZombie, they are able to control their motions and move with purpose almost like a normal human. In *Night of the Living Dead,* they are lumbering without clear, easy motion. *The Walking Dead* presents zombies somewhere in between. Different zombies can have different goals. Some zombies are specifically after brains or are aiming to turn others into zombies, as in *World War Z,* whereas others are seeking specific targets. These are usually the ones that are controlled by a master.

One of the key distinctions between classes of zombies is what causes the zombification and how it is spread. In modern lore, it is usually seen as a neurological disease that is spread through bites. Older zombies, however, were turned by their masters. When looking at the possibility of surviving a zombie apocalypse, it is important to know exactly how zombification may or may not spread. Rarely can zombies be cured. This is usually seen only in older zombie stories, but it is still part of zombie legend. Keeping all these things in mind, let's look at how wights and White Walkers would be classified as zombies and determine whether we can apply the science of zombies to the army of the dead.

In the army of the dead, the White Walkers are the generals and the wights the foot soldiers. This is the master–zombie structure in line with the older zombie paradigm. Though unusual for zombies, it's still well within canon. The wights maintain their ability to fight and, in the case of Benjen Stark, still have some understanding of who they used to be. Any discussion of their motion needs to take this into account. This type of zombieism isn't an infection but a reanimation. Much of the scientific analysis around zombies treats the spread of zombieism as if it were an infectious disease, but that doesn't apply here. Instead, it's more of a hub-and-spoke model, which propagates far more slowly than if wights were able to create new wights on their own. This affects how quickly the Night King can build up his army. In the puppet master model, the goal of the zombies is set by the one that controls them. This is exactly like the setup in *White Zombie*. Just like those zombies, killing the master kills those that he turned. Unlike those zombies, though, wights must be killed in specific ways, and they cannot be cured. Though wights are not the zombie seen in more modern media, they are well within zombie canon and much of the research on zombies applies here.

NEUROLOGY AND BIOLOGY OF ZOMBIES

As zombies took over pop culture, scientists began to contemplate both the possible neurology of a zombie and real-life zombie diseases and parasites. In these cases, they looked at real instances of one organism being controlled by another without realizing it as well as potential causes of standard zombie behavior. The addition of "zombie" to a headline is definitely going to bring in some clicks—take, for example, a recent story about "zombie raccoons" in Ohio. In this case, the raccoons were acting like zombies, walking on their hind legs and foaming at the mouth before falling into a comatose state. With the exception of the last part, that's pretty typical of the modern zombie: zombieism with no master and unusual behavior. The videos, which include other animals such as coyotes and foxes, are really quite shocking. Dr. Tara Smith, a professor of epidemiology at Kent State University, discussed the issue with US national media, ultimately concluding that the animals were suffering from either rabies or distemper.

When wild animals act oddly, most people immediately assume rabies is the culprit. According to the CDC, there are 5,000–6,000 cases of rabies in the United States each year, but typically not more than five in humans (thank goodness). Because rabies spreads through saliva, outbreaks often occur locally. Bats are the most common carrier of rabies in the United States, and most human cases of rabies come from bats. You may not even realize you've been bitten in the event you come in contact with a bat, so if one happens to get into your house, make sure you get rabies shots even if you don't think you need to.

Rabies can be seen as a zombie disease because the virus causes the host to behave in a way that will spread the disease. It causes aggressiveness, which makes the host more likely to bite, and because the virus lives in the salivary glands as well as in the nervous system, the aggressive behavior serves to transfer the disease. The virus enters the peripheral nervous system through a bite and travels through the nerves to the brain. No symptoms are present until the brain is reached, at which point the victim experiences nonspecific symptoms such as headache and fever. Brain swelling soon follows. From there, the disease progresses in two possible ways: furious rabies or dumb (paralytic) rabies. Those descriptions alone make the disease sound exactly like zombieism. The furious form is closer to modern depictions of zombieism, with victims afflicted by aggression, hyperactivity, a fear of water, and the inability to swallow. Death occurs within days of the onset of symptoms, usually from cardiorespiratory arrest. In dumb rabies, the victim is slowly paralyzed and slips into a coma. It's safe to say that if the raccoons were suffering from rabies, it was the furious type. However, rabies doesn't quite explain the raccoons' unusual behavior.[2]

Dr. Smith also suggested that the raccoons could be suffering from a virus called distemper instead. If you have dogs, you know that they are typically vaccinated against both rabies and distemper every year. The distemper virus is different from rabies, but it can also cause neurological problems. Until recently, there were no recorded cases in raccoons, but distemper has been found to spread to a wide range of mammals (but not cats, luckily for them). It is much more infectious than rabies and can be spread through the air. It initially affects the respiratory and lymphatic systems and moves from there to the nervous system. Although the neurological symptoms resemble those of rabies, canine distemper virus operates on the nonnerve cells, or glia, of the central nervous system. The virus destroys a cell's myelin sheath, a protective covering, which causes nerve damage, similar to how multiple sclerosis damages cells. Eventually, the animal's brain swells and they begin to show outward signs of nerve damage. Because the disease affects different parts of the nervous system at different rates in every case, it's hard to say how the disease will progress in a particular case. The destruction of the nerves in this way causes symptoms that mimic zombieism: head bobbing, rhythmic jerking of various muscles (usually the limbs), dilated pupils, watery and wonky eyes, loss of balance, and seizures. It's no wonder people thought the infected animals had a zombie illness.[3] Distemper is the most likely cause of the so-called zombie raccoons, and it causes zombie-like *behavior*, but rabies is ultimately a more zombie-like *disease*. It controls the host's central nervous system and causes them to act in a way that spreads the virus.

There are examples other than rabies in nature that depict the puppet master variety of zombieism. Like the viruses, these zombie creators are trying to further their own reproductive ends. They take over hosts with the aim of creating a situation where they are more likely to reproduce. Technically, this is how the White Walkers use the wights—to kill and leave bodies for the White Walkers to reanimate, increasing the army's numbers and moving the White Walkers closer to their goal of covering the world in ice. A parasite or other animal can invade and use the body of another organism to do its bidding. There are many examples of this in nature, one of which can even occur in humans. In fact, a huge percentage of those reading this book are probably under its spell. Because there are so many examples of parasites that create zombies, I'm only addressing my favorites, all of which operate through slightly different mechanisms.

Wasps are particularly good at controlling other animals. They are known to create zombies out of spiders, cockroaches, and caterpillars. The jewel wasp uses a neurotoxin to make a cockroach do her bidding. The wasp lands on the cockroach and injects a neurotoxin directly into the ganglion, which functions as the roach's brain. The toxin is high in γ-aminobutyric acid, or GABA (more on this in chapter 13's discussion of poisoning), which inhibits the roach's ability to control its muscles. The roach doesn't much care to run away, so it sits there and waits while the wasp builds a nest. The wasp then returns to the roach, eats the tops of its antennae, and then uses them as reins to guide the wasp to the nest. She then attaches an egg to the roach's leg and seals it up. The egg hatches, eats the roach, and the whole thing starts over. The parasitoid wasp does something similar, only to caterpillars instead of roaches. This wasp makes the caterpillar both its incubator and ultimately mother figure. The wasp stings a caterpillar and lays roughly 80 eggs inside. As the eggs develop, they eat the caterpillar from the inside out, making sure to avoid vital organs so the caterpillar stays alive. After bursting out through the skin of the caterpillar, which lives through the ordeal, they build a nest and command the unfortunate host to keep watch over them. After the larvae are fully developed and they have no more use for their protector, the caterpillar dies and the wasps go off to do it all again. (There is a great *National Geographic* video of the process, but be warned: it is *extreme* nightmare fodder.) The mechanism for this zombification remains unclear. What is clear, however, is that nature can be really messed up.[4]

Being from Maryland and having had a small boat when I was a kid, I know all about crabs and barnacles. One is tasty with some Old Bay and washed down with Natty Boh (a favorite Maryland beer); the other is the bane of a boater's existence. I hate barnacles with a passion, but if a crab could hate, its dislike for barnacles would far surpass my own. *Sacculina carcini* is a barnacle that hijacks the green crab's reproductive system and uses it as its own. The barnacle lands on the crab and burrows into it through a hair follicle. From there, it sends tendrils throughout the crab's body that are used to control its behavior. It then pushes out an egg sac and causes the crab's reproductive organs to stop working and atrophy. Instead of trying to reproduce in normal crab fashion, it hangs out and distributes the eggs of the barnacle. Both male and female crabs can be infected and will behave like female crabs. Both male and female crabs treat the eggs of the attached barnacle the same way a female crab would treat her own offspring, with behaviors such as pushing clean water over them and protecting them from predators. The crab stops growing and regenerating its organs, but it can live for up to two years, producing more *S. carcini*. It's pretty impressive that a barnacle can control a crab enough that it thinks it is producing its own children.

One of the most dramatic and well-reported agents of zombification is the fungi of the genus *Ophiocordyceps*. Although it's debatable whether a fungus fully fits the definition of a parasite, in this case it certainly acts like one. It can affect many different types of insects, but the carpenter ant is particularly susceptible. The fungus infects the ant and causes it to climb to a certain height above the forest floor, lock onto a leaf, and stay there until the fungus sprouts a spore out of the ant's head. These then rain down on the other ants below, propagating the fungus. Until 2017, how the fungus was capable of mind control was unclear because the ant's brain did not contain fungal cells. It turns out that the fungus wasn't controlling the ant's brain but rather its muscles. Maridel Fredericksen and her group of researchers at Pennsylvania State University used advanced imaging techniques to look at the cells of infected ants. They discovered that the fungus, which is most prevalent in the head muscles, wraps itself around an ant's muscle cells and creates networks to coordinate motion.[5] These networks direct the ant to a point high enough to spread spores and tell it to lock its mandibles on a leaf until it dies.

There is one type of zombifying organism that affects mammals: toxoplasmosis. About 23% of humans in the United States are affected, but it is more effective at reproducing when it infects mice. This parasitic disease is caused by *Toxoplasma gondii*, a single-celled organism that takes up residence in its host's brain. This protist can only reproduce in the digestive system of cats, and it needs a method to get there. This is where mice come in.[6] When rodents ingest the organism, it forms cysts in the brain that affect the rodent's brain chemistry, taking away the normal aversion to feline urine and making the mouse unafraid of cats. The cat is provided with easy access to lunch and consumes the mouse. This deposits the *T. gondii* in the gut of the cat, where it can reproduce. The organisms then move through the digestive system and are excreted. From there, they mix with soil and the cycle starts over again. Considering the protist causes the rodent to change its behavior, it can be seen as a zombie puppet master. Its control comes from changing the level of certain neurotransmitters, such as dopamine, in the brain. The same organism can inhabit human brains and cause the same changes in neurotransmitter levels. Current research indicates that these changes in neurotransmitters lead to impulsive behavior and may be a trigger for mood disorders and schizophrenia. The brains of people with schizophrenia are three times more likely to contain *T. gondii* than the general population. One study even suggests the culture of a population with widespread *T. gondii* infection can be changed by the protist. This is a weak claim at best, but the concept is still interesting.[7] It's tempting to push these theories and suggest *T. gondii* is responsible for "crazy cat lady syndrome," but there is little evidence for that, thank goodness. (I have two cats. Just two.) Cat owners are just as likely as those without cats to have toxoplasmosis. Still, cats are complicated pets to own and love, so maybe there is something to this theory. Maybe this is why the internet runs on cats . . . but more likely it's because cats are cute. Most zombie-causing organisms rarely have much interaction with humans. It's still good advice to always be careful of zombie organisms, no however interesting you may find them. In the words of the Zombie Research Society, "What you don't know can eat you."

Zombie/Wight Rot

The army of the dead was created by the White Walkers with the purpose of taking over the world. As people died in battle or from other causes, the Night King and his lieutenants (aka the Four Horsemen of the Snow-pocalypse) could reanimate the corpse and add another foot soldier to their army. But did the process of death and decay stop when they were reanimated? Because the wights are made to fight, yet in almost all cases cannot be killed and can't heal, they tend to look a little rough after a fight or two. Normal physical attacks don't have any effect on their ability to stay animated, but what about the bacteria that live on humans and cause decay after death? Would the normal decomposition process be happening while the wights were reanimated? Would it have an effect? It certainly seems from the show that the wights are in various states of injury and decay, but is that based on their state when they died or based on what's happening to their undead yet moving body? The cause of a wight's death, the cold temperatures up north, and the humidity all affect how decomposed a wight might be. Even though a wight hand alone has the strength to strangle a man, it's not going to be as fast or as deadly as a full-bodied wight. Couldn't the Westerosi just maybe stop dying north of the Wall and wait for nature to take its course?

When a body dies, it goes through several stages of decay (four or five depending on who you talk to), including two chemical processes, autolysis and putrefaction. If you are as much of an *NCIS* fan as I am, a lot of these terms will sound familiar. I will admit that though I've watched more than a few police dramas, reading the details of the decay process was a bit stomach turning. Enjoy!

Autolysis is essentially the body digesting itself. The acidity of cells increases as the CO_2 concentration increases. This causes cells to break down and release waste products that eat the cell. The acid in the stomach as well as other enzymes also start to digest the body. Autolysis begins as soon as a body dies. The first, or *fresh*, stage of decomposition starts immediately after death. In addition to autolysis, the body begins moving toward thermal equilibrium, a process called *algor mortis*. In addition, *rigor mortis*, or stiffening of the muscles, occurs due to a chemical change that prevents the muscles from relaxing. Much like Benjen Stark, aka Coldhands, gravity pulls blood to the extremities. There aren't many physical changes in this stage except maybe some blisters on the skin.

As the body uses its oxygen stores, an anaerobic environment is created—the preferred home of gut bacteria, which now move from digesting food to digesting the body. This process is known as putrefaction. The gas released from putrefaction leads to the second stage, bloat. This is exactly what it sounds like. The gases cause the body to distend as bacteria liquefy its contents. This causes the body to swell. If enough pressure accumulates, the body will rupture, forcing out the frothy liquid. I think this is probably the grossest stage of death.

After bloat, the body enters active decay, which involves the feeding of maggots and more liquefaction of the body through putrefaction. I won't talk much about insects because there is probably only one species of insect north of the Wall, the chironomid midge, and it isn't involved in the decay process. If potential wights were further south, however, this is the point where bugs would really get involved. This stage ends when even the maggots have nothing left to feed on. At this point, most body mass is gone and, for those who subscribe to the five-stage model of decomposition, the body enters advanced decay. There's not a ton left to break down and fewer insects are interested, so decay slows. Many roll this stage into the active decay stage. The final stage is skeletonization. This is exactly what it sounds like. At this point the body is dry and potentially just bones. Decomposition has essentially ended.

There are many factors that affect how a body decomposes: how exposed it is, availability of oxygen, humidity, cause of death, depth of burial and soil type, presence of insects and bacteria, clothing, and, most important to this discussion, temperature. In the case of the army of the dead, they are lucky when it comes to decomposition. The climate is cold, most likely dry, and has very few insects. Since there are effectively no insects involved, the decay would be caused by autolysis and bacteria. Unfortunately—or maybe fortunately, if you are the Night King—bacteria need water to live, and enzymes and bacteria can't function below a certain temperature threshold. The chemical reactions that cause these processes need energy, in the form of heat, to start and continue. Without heat, they can't do their job, so the body would be left in pretty good condition. The blood would still pool in the direction of gravity and rigor mortis would still set in; however, the rest of the decay process essentially halts below a certain temperature. Coldhands is a pretty fair representation of a reanimated corpse, though he would have to be strong to overcome the effects of rigor mortis. In cases where someone dies in extremely cold condition, they become mummies instead of decayed corpses. A good example of this are the bodies left on Mount Everest. It is usually not possible to recover the body of a lost climber on Everest in the "death zone" without risking one's own life. Of the 216 people who have died while trying to summit, only 66 bodies have been recovered. The other 150 have been left to mummify in the cold and dry conditions. In several cases, the bodies trail markers to climbers on their way to the top. George Mallory, whose remains were found 75 years after his death, was virtually free of decomposition. Several other notable bodies are that of Hannelore Schmatz, nicknamed "the German woman," who died close to Camp IV and remained looking untouched until the wind blew her remains away. One area near the summit is known as the "rainbow valley," named for the brightly colored, undecayed parkas worn by the corpses on the slope. "Green Boots," an unidentified climber whose corpse is also in pristine condition, lies in a small cave very close to the summit, right beside the path taken by those attempting to reach the top. What this tells us is that when it comes to the army of the dead, even if you assume a violent death and long exposure to the elements, the wight should actually be less decomposed than it appears—assuming, of course, that it can be reanimated in the first place and that reanimation can overcome rigor. It already has to overcome a whole lot of other biology, so I guess it can also overcome rigor.

Zombie Neurology: What's Going on in Their Heads?

Until now, I've talked about real-life zombies and many examples in nature of one species controlling another. In the case of White Walkers and wights, I hate to say it, but they aren't real. This hasn't stopped scientists from asking the question, "But what if they were? What type of disease might they have?" Neuroscientists Timothy Verstynen and Bradley Voytek wrote a whole book about this. I won't go into too much detail, but I'll pull out the highlights that might apply to wights specifically. In *Do Zombies Dream of Undead Sheep?*, the pair investigate the potential cause of traditional zombie diseases with symptoms of lumbering gait, lack of language production and comprehension, anger, inability to feel pain, and hunger for brains.[8] They even coined a term for zombie disease: Consciousness-Deficit Hypoactivity Disorder (CDHD). In the case of wights they only exhibit a small subset of CDHD symptoms. They can't talk, but they are still able to process instructions from a White Walker. They are still very coordinated and are able to move quickly, and they're pretty deadly with a sword. They are definitely angry and ready to kill, but they aren't able to feel pain and they don't seem particularly hungry. Fueled by rage and lacking both the power of speech and the ability to feel pain, their brains—or what's left of them—are less damaged than those of other zombies. However, there is still some specific damage.

You have probably used the term "lizard brain" or "reptile brain" at some point when referring to an instinctive reaction or impulse. The other day, as I screamed and ran away from a tiny spider, I made a reference to my "lizard brain," meaning the part of my consciousness operating illogically, on instinct alone—in this case, fueled by my fear of spiders. (In my brain's defense, no creature needs that many legs. Anything more than four can only be used for evil.) When people talk about their reptile brain, what they are really talking about is the fundamental parts of the brain that are in charge of the things we do on instinct. The amygdala (which is plural, like "data") are two bundles of nerves behind your temples. These nerve bundles regulate emotions and initiate reactions such as the fight-or-flight response. The *hypothalamus*, another nerve cluster, regulates functions including sleep, hunger, and response to stressors and also plays a key role in the fight-or-flight response. Verstynen and Voytek theorize that a zombie has an overactive fight-or-flight response with a bias toward the fight instinct. When a human encounters a stressful situation, the amygdala jumps into action. In the case of my spider mishap, my amygdala saw a serious, base-level threat on eight legs headed my way. They triggered my hypothalamus to release a hormone that told my pituitary gland to step up. The pituitary gland released another hormone that told my body to go into stress response mode. With that, my adrenal glands produced adrenaline and set my body on edge, ready to squish the spider or run for the hills. These glands also produce testosterone and cortisol. The result is that my lizard brain—although it is technically a subset of the *limbic system*—decided that my best chance to live would be to scream like a 5-year-old and flee the scene at a speed I will never reach again. If fleeing wasn't an option, however, the same hormones would amp up my body to fight. At this point, all higher-level thinking is turned off and I'd be running on pure instinct. Essentially, my amygdala have hijacked my brain and turned me into a deadly spider assassin (or, rather, Usain Bolt).

According to Verstynen and Voytek, the amygdala of zombies are always in control, and the *neocortex*—the part of the brain that actually likes to think—is on vacation. Zombies, they posit, have adrenaline, testosterone, and cortisol constantly coursing through their veins at very high levels. This amplifies their arousal and heightens their fighting abilities. Generally, the *orbitofrontal cortex* is in charge of keeping the amygdala in check. It's needed for higher-level thinking such as strategizing for a battle. Wights and zombies don't have the ability to coordinate or plan as a group, however, so it wouldn't be surprising if this part of the brain were missing or severely damaged. If wights had sustained the same type of damage to this region of the brain in addition to their constantly engaged fight-or-flight response, they would exhibit all of the behaviors Jon and the Suicide Squad witnessed in "Beyond the Wall."

Both wights and zombies are able to understand and respond to sound. Seeing as the wights respond to the noises of humans, they have to be able to hear and process at some level. Their ears, or what's left of them, as well as their brain stem and *primary auditory cortex* must be in decent working order. The signals from their ears are relayed through these parts of the brain to determine where a sound is coming from. That is all a zombie or wight would need to be able to act like a zombie. They need only the ability to process that there is a sound and figure out where it's coming from so they can head in that direction. In most humans, the next stop in language processing is *Wernicke's area*. This is where the brain turns the sounds back into language. When this area is disrupted and someone can no longer understand language, it's called *Wernicke's aphasia*. This area may or may not be functioning in wights, but it is definitely out of commission in most zombies. In the case of wights, it's a little more unclear. Wights listen to their White Walker masters. There has to be some method of communication, even if it hasn't yet been made clear or if we are supposed to assume it's magic. Regardless of how a White Walker is "talking" to a wight, there needs to be some function in Wernicke's area to turn those signals into action. Having a brain that is still intact enough to process commands makes wights stand out from the normal zombie crowd. The complement to Wernicke's area is the *Broca area*. This part of the brain is responsible for speech and language output. It turns what the brain wants to say into movements of the mouth and vocal cords to produce coherent language. Neither zombies nor wights have that ability. In fact, the White Walkers aren't big talkers either. When the Broca area is damaged and language can't be produced, it's called, as you can probably guess, *Broca's aphasia*. It's fairly clear that most zombies, all wights, and potentially the White Walkers themselves suffer from Broca's aphasia. In general, zombies and wights have similar brain damage; however, given their ability to process instructions from White Walkers, it would seem that wights don't suffer from Wernicke's aphasia.

Zombie Statistics and a Survival Plan: Can Westeros Get Out Alive?

Zombie outbreaks are quite different from what we observe of White Walkers and wights because zombies typically originate from an infection scenario rather than a puppet master arrangement, but the spread is similar to some degree. Wights kill and the dead are reanimated by the White Walkers to attack the living. From a mathematical modeling point of view, the two outcomes don't differ much, assuming everyone killed is then reanimated by a White Walker. Given the size of their army, this isn't a bad assumption to make. The main difference is that killing a White Walker will take out all of the wights he'd created. The zombie spread is similar, but the fight is different. Only a few people in *Game of Thrones* possess the ability and weaponry to kill a White Walker, so the survival of others north of the Wall is dependent on the same skills as surviving a more traditional infectious disease model. Luckily, many professors and grad students with access to fast computers like to do such things in their spare time. There are 37 papers about zombies on arxiv.org, the academic preprint server. One in particular, published in 2015, is the most comprehensive statistical analysis of the potential spread of zombieism. It is fairly general and can be easily modified to fit the wights of *Game of Thrones*. Alexander Alemi and his colleagues at Cornell University started with a traditional disease spread model, the "Susceptible–Infected–Recovered" (SIR) model, and generalized it to the SZR model, "Susceptible–Zombie–Removed," where "removed" means a zombie has been killed (for good).[9] This model looks at the change over time in these three populations based on parameters set at the start. The kill parameter gives the probability that a human kills a zombie, and the bite parameter indicates how likely it is that a zombie will bite a human, thereby infecting them. In the case of the wights and White Walkers, it's not the bite that changes a human, it's the kill. When looking at how this model applies to *Game of Thrones*, the bite parameter can be easily changed to the wight parameter. (It even rhymes!) Alemi's paper deals directly with zombie outbreaks in different sized populations. Considering that wights had small populations to work with until the end of season 7, this makes the paper's model applicable. They used a simulation that models one-on-one interactions between zombies and humans, using probabilities of bites (or wights) and kills (of wights by humans). They found that the probability of stopping a zombie outbreak is directly proportional to the ratio between a human's ability to kill a zombie and a zombie's ability to kill a human: kill/wight. If that ratio is 1:1, meaning they are equally likely to kill each other, humans will always stop the apocalypse. Their model showed that one side will always win: either humans kill all the zombies, or the zombies turn all the humans. As the war rages, either everyone will be turned, or all zombies (or wights) will be killed; however, the results were somewhat dependent on population. If there are more than 100 people, they are as likely to survive as 100,000 people. If there are fewer than 100 people, however, there is some fluctuation, and if there are fewer than 10 people, no one's getting out alive in most cases. These cases all involve only one zombie thrown into a population of humans. These cases were applicable when the Night King was first created or if a wight were to infiltrate a wildling camp. Let it be said that in the current state of affairs in Westeros, we are *well* past that point.

This model only gives the outcome for an evenly distributed population in which a single initial zombie can interact with any other person. This is also not the case in the North. The population north of the Wall is not evenly distributed but clustered in groups of travelers, be they wildlings or ranging parties. Alemi's team modeled this scenario as well by creating a grid with one person per box and dropping one zombie in the mix. It can only turn neighbors, and its success rate is governed by the same kill/wight ratio from the first model. What they found was that zombieism spread in a network. All isn't doom and gloom, however—the good news here is that there are pockets of survivors. There is a critical threshold of interactions that lead to bites, under which the apocalypse is halted but over which everyone is doomed. If humans are slightly less than half as good at killing zombies as zombies are at killing humans, humans will stop the spread of zombieism. Again, this is all assuming one zombie and a population limited to interactions with neighbors. The final map of infection leads to a fractal pattern. This isn't really relevant to Westeros but is interesting nonetheless. From this initial model, we can then move to a more realistic model of a population that is spread out in clusters, as seen in the United States. In this model, zombies were also given the ability to move, however slowly. Humans were not given the ability to move, under the assumption that mass hysteria would lead to road closings. (I'd say that's a fair assumption.) They assumed one zombie for every million people. What they found was that most of the US population became zombies within a week and the spread depended largely on population density. Remote areas resisted the fall into zombieism well past the four-month mark. After running their simulation a number of times with different starting points, they created a zombie susceptibility map. If there is a zombie outbreak, you should definitely avoid major cities, but more importantly, you should avoid the connections between cities; the pathways between cities can be affected by either city, so you are more likely to be attacked in those areas.

What does this mean for Jon, the wildlings, and really, at this point, the rest of Westeros? It's hard to say exactly, but I think they are pretty much screwed. The key difference between the zombie model presented here and the army of the dead is that there are nodes—that is, White Walkers—that can negate the entire model. According to Dany, there are roughly 1,000 White Walkers. If we assume some of the Westerosi—that is, an army armed with dragonglass and Valyrian steel that is evenly matched with Walkers—are able to kill White Walkers, half the army would be lost in each direct battle with a Walker. It would take an army of 2,000 evenly matched and armed soldiers to have a statistically probable chance of killing the Walkers. That, of course, assumes they can fight through the wights to battle the Walkers. Dany also estimated there are about 100,000 wights, so each dragonglass-armed soldier must kill 100 wights to then have a 50% chance of killing a Walker. Meanwhile, the zombie plague has spread across nearly the entire continent in a week or so, stopping Jon and Dany from recruiting the needed soldiers and adding to the opposing force of wights. So, the odds are amplified against our heroes as the battle progresses. At this point, if you are a random resident of Westeros, do what you would do anyway: get off the road, get out of the city, find somewhere with fresh water and food, and just try to wait it out. Really, though, there's not much hope for you barring the deus ex machina of Drogon (and we saw how that went in season 7).

Zombies so often are used as a dramatic tool to bring out the true nature of each main character. In *Night of the Living Dead*, the true horror was the collapse of the group trying to survive, and the zombies served to push the collective to self-destruct. In Max Brooks's *World War Z*, the story uses a zombie apocalypse to provide commentary on the modern geopolitical landscape. In *Game of Thrones*, the army of the dead serves as a monster MacGuffin to reinforce exactly how altruistic or short-sighted (or both) our human characters are. In that way, although the rules are different and the statistics may not apply in the same way, the army of the dead serves as a dramatic vehicle driving the story, just as zombies do.

Bonus: Zombie Dragons

As discussed in chapter 3, the season 7 episode "Beyond the Wall" left a number of unanswered questions. I addressed some of them but saved one of the big questions for this chapter: How did the wights pull zombie

Viserion out of a lake, and where did those chains come from? We need to make a few estimates to get to the answer. First, how long would it take for the ice to freeze solid enough to hold a dragon? Second, how long does it take to forge or find and attach chains to Viserion? Third, how strong would the army of the dead need to be to make this happen, and how would they be able to pull hard enough without slipping on the ice? As for how thick the ice needs to be to hold a dragon, there's a pretty straightforward equation for that. The required thickness of the ice is dependent on the square root of the dragon's weight. (Technically, this equation was derived for planes . . . but same thing, really.)[10]

$$S = C\sqrt{W},$$

where S is the thickness of ice needed in inches, C is a constant that depends on the type of ice—either sea ice, river ice, or lake ice—and W is the weight of the plane (or dragon) in tons. This formula works only for ice created at about 16°F; for higher temperatures, S will be about 25% thicker. Since Viserion fell in a lake, $C = 3.75$. Assuming Viserion is about the weight of a Boeing 747 in tons, $W = 200$ tons. These values indicate that 4–5 feet of ice would be needed. At about 20°F the lake would freeze about 1 inch per day.[11] This would give the White Walkers about two months to procure an insane amount of chain. Luckily, due to the conditions, Viserion wouldn't have decayed much by then.

As for the chains, if you type "where did the Night King" into Google, the second suggestion, after ". . . come from?" is ". . . get those chains?" Clearly this is something everyone wants to know. Some have suggested they are of giant origin. This would certainly explain how they were forged, since I don't think wights like forges. There's no great scientific explanation for how creatures that clearly can't swim found and attached giant chains to a dragon in a lake, so I'm going to leave that as an exercise for the reader. However, some have asked how a 200-ton dragon could be pulled out of a lake without ripping his wings off. This is a valid question; that amount of stress is really too much for most joints to handle. As I'll elaborate further in chapter 7, however, dragons most likely have extra-strong bones, which would counteract the wing-ripping problem. (Turns out it doesn't really matter.)

As for how strong a wight has to be to pull those chains, the first thing we need to know is how heavy Viserion would be when submerged in water if buoyancy is considered. I've assumed Viserion weighs about 200 tons, or 400,000 pounds, so the downward force is ~400,000 pounds. Buoyancy would hold him up a bit and is given by the equation

$$F_B = \rho g V,$$

where ρ is the density of water and V is Viserion's volume. Assuming again that he's roughly the weight of a 747, his volume is about 876 m^3. This equation gives a surprising answer: if fully submerged, Viserion would be subject to a buoyant force of about 2 million pounds. This is five times his weight. He would float! Guess that answers the question of how the wights swam down to attach the chains.

The question, then, is how hard the wights would have to work to drag Viserion close enough for the Night King to reanimate him. In reality, this isn't much of a problem, either. He was already above the water, so the Night King could have touched his snout without much need for pulling anything besides his head. If Weiss and Benioff had simply paid attention in physics class, none of these pesky plot holes that the internet forums can't stop talking about would even exist.

5

Regular Steel, Made in Pittsburgh

Sharp steel and strong arms rule this world, don't ever believe any different.

—Sandor "The Hound" Clegane, *A Clash of Kings*

Many scientific advances have been made through efforts to create tools and weapons. Metallurgy is no exception. Copper and iron were refined to be forged into weapons and tools and were later mixed with tin and carbon, respectively, to make bronze and steel, alloys even better suited for metalworking. In fact, steel is almost synonymous with weaponry. Imagery of the sun glinting off a knight's plate armor or the clang of swords clashing in battle is a hallmark of most novels set in the Middle Ages. *Game of Thrones* is no exception. The knights of Westeros are known for their swords, which are handed down to their children as one of their most valuable assets. Outside of fantasy worlds, steel is no less important. The transition from stone tools and weapons to copper and bronze was a huge step in the advancement of civilization. The discovery of iron was another step. The discovery that mixing iron with carbon would create steel launched humanity into the modern age. Steel is durable and the iron from which it's made is relatively easy to smelt and forge. Steel made the weapons of the Middle Ages and the skyscrapers of today. But it's not infallible—just ask Jack and Rose of *Titanic* fame. If you're gonna send hundreds of people into the icy North Sea in a tub of steel, ya better get it right. It took humans a long time to figure out the best way to make quality steel and engineer it for a variety of applications, and even now the method is still being perfected.

In Westeros, there are two types of steel used to forge weapons: normal steel and Valyrian steel. I'll talk about both, but this chapter will focus on normal steel and its creation and uses throughout history. The microscopic reasons for its macroscopic properties are surprisingly similar to those of the ice wall I talked about in chapter 2. Steel also has its limits, as many a Brother of the Night's Watch found out when fighting White Walkers (as did Jack and Rose). Steel works well at normal temperatures, but it can be negatively affected by cold. Hopefully by the end of this chapter, I'll be able to explain why steel just can't cut it (ha!) against the army of the dead. In later chapters, I'll talk about how Valyrian steel and dragonglass can do the jobs that normal steel cannot. Now, let's take a little materials science tour through human history.

Hard, Soft, Brittle, and Bendy: Why Steel?

Steel revolutionized warfare throughout the world. Before we can really talk about why it had such a profound effect, though, we first need to understand what it actually is. Really, it's worth defining metals and alloys and learning what makes them so well suited for weaponry. Before copper was discovered and isolated, weapons and tools were made of stone and wood. Although these certainly did a decent job, they were easily broken or chipped and were not able to hold an edge. In short, they were brittle or soft. Many metals are also chemical elements, meaning they're made up of one type of atom and are found in nature. Metals make up the majority of the periodic table. You could probably list most of their properties if asked at pub trivia. To be classified as a metal, an element must have several properties. It must be shiny, hard and opaque when solid, and, with the exception of mercury, solid at room temperature. Metals are also ductile, meaning they can be drawn into a wire, and malleable, meaning they can be hammered into a shape. In addition, they are good conductors of heat and electricity. Many of these properties make them well suited for weaponry, but a few questions remain: How are metals structured on a molecular level? How do they fail? And how can they be made better?

The key to most of a metal's unique properties is how its atoms are bonded. This bond, not-so-profoundly known as a *metallic bond,* is different from other types of atomic bonds and occurs only in metals. In a metallic bond, the outer electrons of the atom are only mildly interested in staying put. A metal is more easily thought of as a set of fixed, positive ions immersed in a sea of electrons. For an element to be electrically conductive (that is, to convey moving charges), there need to be charges to move, and those are provided by a metallic bond. In some cases, this means the movement of ions, but in most cases, *current* refers to moving electrons. In insulators (a different material type with physical properties opposite to those of metals), there are very few electrons that are comfortable with roaming freely. Metals, however, have a cloud of free electrons wandering about, so giving them a push in the form of voltage causes them to move and conduct a current. This is true for heat, too, because heat is transferred by moving particles bumping into other particles and making them move. (If you haven't yet had the chance to read chapter 3, it explains heat transfer more elegantly.) This sea of electrons is also why metals are shiny. When light hits an object, some of that light is absorbed, some is scattered, and some is reflected. All of this depends on how the electrons are arranged in the atoms and molecules that the light is reflecting off of. In the case of nonmetals, the electrons are stuck to the atoms and can only move in certain ways, meaning only certain colors of light are reflected. In the case of metals, the electrons can move in many different ways and can therefore reflect many different wavelengths of light. They can't always reflect all colors, which is why different metals appear to be different colors, but in general, metals reflect a broad spectrum of light.

As I mentioned in the discussion of ice in chapter 2, solids have different crystal structures. In the case of metals, they have one of three different structures: body-centered cubic (BCC), face-centered cubic (FCC), or hexagonal close-packed (HCP). HCP looks a lot like a slightly offset honeycomb stacked on top of another honeycomb. (I'll address this structure in more detail in chapter 6.) The two metals most relevant to this discussion are copper and iron. Copper has FCC structure, which looks like a cube with an atom at each corner and another atom in the middle of each face. Iron can have either FCC or BCC structure depending on how it was cooled and what impurities it may have. BCC looks like a cube with one atom at each corner and one in the center of the whole cube (figure 5.1).

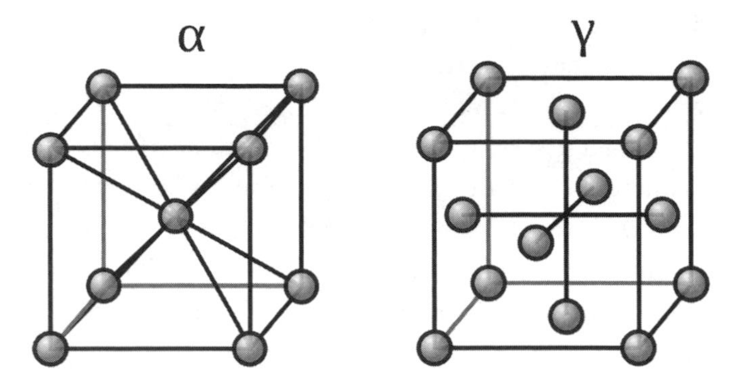

FIGURE 5.1
Left, body-centered cubic (BCC) structure; right, face-centered cubic (FCC) structure. The BCC structure is called austenite.

Because these atoms are held together with a metallic bond, they can roll over one another without having to break and reform bonds. If the metal is pushed a little bit, it will go back to the way it was before, but if it is really hammered, the atoms will move around and get comfortable somewhere else without breaking the entire structure apart. The bonds don't have to break to allow the atoms to move. This gives a metal its ductility and malleability. Just like the crystal structure of the ice used in the Wall, a metal's crystal structure isn't perfect throughout, and that is what gives a metal its hardness. A *crystal grain* is a section of well-aligned atoms. The boundary between two crystal grains, known as the *grain boundary*, is considered a defect. Within a crystal grain it is easy for the atoms to roll over each other, but it isn't easy for them to move near grain boundaries because the structure isn't as well ordered. Because the atoms can move locally but not over long ranges, the material is malleable yet still hard. The more a metal is worked, the more grain boundaries are introduced and the harder the metal becomes. This is one reason iron is worked many times over. This hardness comes at a cost, however. As seen in ice, defects and boundaries are the lines along which a material can break.

When talking about metals, it's important to define what it really means for a substance to be *hard*. It's not the same as being *strong* or *tough*, which are also potential properties of solids. A hard material resists being scratched or dented. Stone or wood, for example, can be scratched or dented relatively easily, but it takes a lot more force to scratch or dent a metal. The hardness of a metal is measured on the Mohs scale, which you may have heard of in relation to gemstones. The scale ranks materials so that anything with a lower number can be scratched by anything with a higher number. Diamond, with a hardness of 10, can scratch anything, but lead (1.5) will be scratched by almost anything it comes in contact with. Copper has a hardness of 3, both steel and iron are roughly 4.5 on the scale, and tungsten carbide is 8.5–9.

Unfortunately for metals, being hard isn't the same thing as being tough. Brittleness is defined as how easily something breaks without first deforming. Play-Doh isn't hard, but it isn't brittle, either. It will deform easily before breaking when force is applied. Toughness is the opposite of brittleness. If something is tough, it will deform quite a bit before it cracks. Therefore, Play-Doh is soft yet tough—kind of like a character in a young adult novel.

Strength, which is different from hardness or toughness, is the ability of a material to withstand applied force without breaking or bending. To determine a material's strength, you look at how it behaves under stress, which is a force per unit area. There are three different components to stress: compressive stress, tensile stress, and shear stress. Materials don't always have the same strength under each type of stress. A material may be very strong when pulled (tensile stress) but not so strong when pushed (compressive stress) or subjected to sliding forces (shear stress). As with its other properties, a metal's strength under different stresses comes from its metallic bond. These bonds really do not want to break, so it takes a lot of effort to pull the atoms apart. Because of this, metals can withstand a lot of force before failing. Different metals have different properties depending on how many electrons each atom can contribute to the electron cloud. Though it's counterintuitive, atoms with fewer electrons to donate have a higher conductivity than those with more. The low number of electrons in the cloud allows for motion with little resistance. It does, however, make the metal softer and more malleable. Iron has more electrons in its cloud, which makes it harder for the atoms to move and eventually separate. Iron is therefore stronger than gold or copper but less conductive. You can try this for yourself if you have aluminum and cast-iron pots; you'll find that the aluminum heats up much more quickly, but it's easier to dent than cast iron.

I've made it sound like metallic bonds are the best thing ever, but there are definitely some downsides. Because metals have electrons that are out in the open, they bond easily with other atoms. With all those electrons jumping around, it's really easy for them to bond with atoms near them. This is why metals rust. The iron ions readily combine with oxygen in the presence of water and create iron oxide, or rust. The patina that appears on copper and other metals is just their form of rust. These are examples of *oxidation*, a reaction similar to the combustion reaction I talk about in chapter 9. The electrochemical properties of salt speed up oxidation and make things rust more quickly. For those of you in snowy areas, this is why you want to get your car washed regularly if you've been driving on roads covered in salt after a snowstorm. The paint on your car is an excellent barrier, but if you are anything like me, the paint has a few scratches. In the case of iron, rust makes it easier for water and oxygen to bond with more iron atoms and create even more rust, which gradually destroys the strength and other desired characteristics of iron. Metals like copper are lucky. The same reaction occurs, but the patina covers the metal so completely that it is protective rather than corrosive. Copper oxide is green, which is why copper and bronze (an alloy of copper and tin) turn green out in the elements. As I'll explain further in the fire chapters, copper burns green for the same reason it turns green: the reactions are the same. That metals are so reactive also means they are rarely found in their elemental (pure) form in nature. They are normally found as an ore, a compound of the metal and other elements. To convert ore into something useful, it must go through the smelting process.

Isolating Metals: Smelting and the Dawn of the Bronze Age

Smelting is a process that separates a metal from the rest of the ore. Not all metals are found in ore—gold is one notable exception—but most are. If gold were found in ore, there would be far less drama in those old Western films; smelting isn't nearly as exciting as panning. Usually, an ore is a compound of oxygen, sulfur, silicon, and the metal you are trying to extract. The first step in smelting is called *roasting*, and it's pretty much exactly what it sounds like. Like roasting a turkey, the ore is heated to a relatively low temperature (below the melting point) and kept there for a while. At this point, the ore undergoes a chemical reaction due to thermal decomposition. The unwanted parts of the ore (usually sulfur, silicon, hydroxide, or carbonate) break their bonds with the metal and recombine with each other and the oxygen and water in the air. They leave behind the metal with some oxygen attached.

Next, that oxygen needs to be removed from the metal. A *reduction reaction* is the general term for a chemical reaction that removes oxygen. It's the chemical opposite of oxidation, which we covered in the last section. These two reactions are known by the combined term *redox reactions*. In the case of a metal oxide, it needs to be put in an environment where the oxygen wants to leave the metal and bind to something else. Traditionally, the metal oxide was placed in a high-temperature environment with little oxygen and lots of carbon monoxide (CO) from the fire used to heat the metal. It's essential that this environment have almost no oxygen but lots of CO, which would really rather be carbon dioxide. Luckily for it, there's some oxygen that's looking for a new home. The oxygen in the ore jumps off and joins with CO to become CO_2. Other compounds known as *fluxes* can be used to speed up, or *catalyze*, this reaction. Fluxes also bind with unwanted compounds to get them out of the way. Quicklime is a very common flux that reacts to remove silica, phosphorus, and sulfur. After the redox reaction the metal is left in its elemental state. Congratulations, you now have iron. Just iron.

But the metal is actually not the only thing left. The byproducts of the process—the impurities that were removed and the products of the reactions—are still around. These unwanted byproducts are called *slag*. Normally, byproducts are a bad thing. In the case of iron smelting, however, slag serves a secondary purpose. Iron really likes to bind to oxygen, so the slag covers the molten iron to prevent it from gaining back the oxygen it just lost. The slag is then easily removed, cooled, and turned into granules. These granules are either disposed of or used. There is a whole industry devoted to using and disposing of slag. The National Slag Association was founded in 1918 and refers to slag as the ultimate sustainable product. According to them, slag has many, many industrial uses. It is used mainly in concrete and cement, but certain types can be used in asphalt and agricultural applications as well. In many cases, though, slag is simply disposed of.[1] Roughly 7 million tons of slag are produced each year. The majority of it is sent to landfills. The granules are mixed with water to create a slurry, which is pumped into landfills. This poses a huge problem because the mixture is very alkaline and thus ruins the soil. Alternative methods for slag cleanup are being researched, but it is very difficult to dispose of safely. When you hear political arguments about regulating the disposal of waste products from steel manufacturing (which starts with iron smelting), slag is one of those waste products, another being carbon dioxide (the major greenhouse gas).

THE BRONZE AND IRON AGES

It was the process of smelting that took metal from being nothing more than shiny flecks in a chunk of rock to being the gold standard for creating weapons and tools. The copper smelting process ushered in the Bronze Age and later the Iron Age, both important eras in human history. The first metals smelted were lead and tin due to their low melting temperature, but both were too soft to be used in weapons or tools. Bronze is an alloy, or a mixture, of copper and tin. Tin had already been isolated, so the key to the process was copper. The smelting process I explained in the last section describes the modern smelting process. In the Bronze Age, it was difficult to create fires that could burn hot enough to bring ore to the temperatures that would start the chemical reactions involved in smelting. It is theorized that copper was first isolated from ore by accident around 5000 BC using a pottery kiln. Kilns are able to reach extremely high temperatures, much higher than the fire in them. A kiln needs four key elements to operate effectively: insulation, a way to get stuff in and out, oxygen, and fuel. I'll talk more about the combustion reactions that create heat in chapters 9 and 10. For now, the important thing to know is that oxygen causes a fuel to burn more quickly and more intensely, creating more heat. This is why people blow on fires when they are trying to light them. In a kiln, there is usually burning wood or coals inside an insulated chamber. The chamber keeps the heat in just like an oven and allows temperatures to rise above those produced by the fire. Oxygen flows in through a vent at the bottom of the kiln and out through a chimney via natural convection, which makes the air rise and carries the heat of the fire up with it. Wood burns at about 600°C, but a kiln can reach temperatures as high as 950°C, the approximate temperature needed for the copper smelting process. Air flow can be controlled with the chimney flue, but the process was refined after the invention of the bellows, which allow better control over the airflow.

As far as metals go, copper is good at a lot of things. The crystal structure of a copper atom is FCC, and it has one electron to give to the electron cloud. The electron cloud density means it binds easily and quickly with oxygen, forming a barrier that stops corrosion. The crystal structure means that atoms can slip over each other in the maximum number of directions, which is what makes copper so ductile. In some cases, copper pipe can even be bent by hand. The ease of slip means that copper is also quite tough; it will dent or deform before failing. The tradeoff, however, is that copper isn't very hard. Although copper was better for tools and weapons than stone, lead, or tin, it still wasn't great. No one wants a sword or hammer that is easy to bend. It's great for thermal and electrical conductivity and was (and still is) used for cooking, but electrical conductivity didn't interest people all that much back then.

How it was discovered is unclear, but around 4000 BC, people found that mixing tin with copper created a metal that was much better suited for tools and weapons than copper alone: bronze. Bronze was so useful that it has a whole historic age named after it. Bronze was much harder than other metals of the time, it was tough yet easy to work with, and it wouldn't corrode. These properties come from the arrangement of the tin atoms within the copper structure. There are three different ways the tin atoms might fit in. Most modern bronze is composed of 88% copper and 12% tin, but ancient bronze contained anywhere from 6% to 20% tin, with different percentages giving different properties. The incorporation of the tin atoms is dependent on how the metal is cooled and worked. One way tin is incorporated is by substituting it for copper atoms in the FCC structure. This type of incorporation happens when a mixture containing <11% tin is cooled slowly. This gives the copper and tin atoms time to arrange themselves into the ordered FCC lattice, with some tin atoms pretending to be copper ones. Tin atoms, number 50 on the periodic table, are much bigger than copper atoms, back at 29. Tin's larger atoms cause strain when they are jammed into places they shouldn't fit. I like to think of them as Baymax from *Big Hero 6* trying to stuff himself into a grid of BB-8s. The strain makes the new alloy much stronger than copper on its own. This same idea of having larger atoms or ions shoved between smaller ones is the same principle used in strengthening Gorilla Glass. Because the atoms in the alloy, whatever type they are, still sit in an ordered way they are still able to slip over each other and the alloy still retains the ductility of copper. At 11%, roughly 1 in every 10 atoms will be tin and 9 will be copper. This is enough that each little FCC box should have at least one atom of tin in it. If the percentage of tin goes any higher than that, the atoms don't always fit neatly in the FCC lattice. At this point, islands containing much higher concentrations of tin can form as the copper cools. Cooling the mixture quickly makes these islands more likely to form. The higher the percentage of tin, the more islands there are. As with pykrete in chapter 2, these islands stop slippage. This makes the alloy much less ductile but much harder. The fancy name for these tin-heavy roadblocks is *engineered discontinuities*. Changing percentages and cooling rates can control many of the macroscopic properties of the resulting bronze. If you want to know more about copper and its alloys, I would highly recommend visiting the Copper Development Association's website (https://www.copper.org); they have information on just about everything you might want to know about copper.

You can see, then, how the discovery of bronze revolutionized the world. To move ahead as a civilization, you need to be able to make stuff and kill the people who are trying to kill you. You need tools and weapons. Unlike stone and the elemental metals known at the time, bronze could be both cast (meaning melted and formed using a mold) and forged (meaning heated and hammered). Bronze was strong enough to be used for objects that must keep their integrity through lots of wear and tear, such as boat fittings, hoes and sickles, cookware, and saws. Bronze was also used for money and art, which . . . well, that breaks the tools and weapons theme, but still. Of course, bronze was also used for swords. Bronze could be made with different amounts of tin to better suit different types of weaponry. Shields and armor, for example, required bronze to be hammered into sheets that hopefully wouldn't sustain repeated blows, so they were often made of bronze with a lower tin content. Lower tin content means the atoms can slip over each other more easily, making beating out a shield or helmet a lot easier. It also makes the bronze less brittle. Probably a good thing when you'd much rather a shield dent than give way altogether; not so much for a sword, though. Bronze used in swordmaking often had a higher tin content. The additional tin made the alloy harder and the sword better able to keep its edge. There is, however, one really big downside to making everything out of bronze: availability. Neither copper nor tin is super abundant in nature. To make matters worse, copper and tin deposits are often not found near each other. This problem spurred a lot of trade in the ancient world, which also meant that any disruption to the trade routes caused difficulties in making and using bronze. There is some very interesting reading out there about the fall of the Bronze Age, but considering this chapter is about steel and physics, I won't go into details here. Suffice it to say, as bronze became more expensive and harder to acquire, advances in ironworking took off, ushering in the Iron Age. (Historians, please excuse my gross oversimplification; it really is fascinating.)

The Iron Age, which probably should have been called the Steel Age, started between 1200 and 1100 BC. I'm not an archaeologist, so I'm sure there were reasons for calling it the Iron Age—I'm just not clear on what they are. People were working with iron during the Bronze Age, too, but the work wasn't advanced enough to create tools and weapons that could rival those of bronze. With the components of bronze in limited supply, people were forced to get better at working with iron, and they stepped up to the challenge. The hurdle in creating steel is iron's high melting temperature of 1538°C. Kilns of the day could not reach the temperatures needed for iron smelting, so people got creative. One of the clearest explanations of ancient iron smelting is from the peer-reviewed Online Reference Book for Medieval Studies (https://the-orb.arlima.net/index.html). Their tech is about as ancient as their subject matter, but it is a wonderful resource for those interested in medieval metallurgy.

Iron was first smelted in a *bloomery*, which looks a whole lot like a kiln. Charcoal, which burns hotter than wood, was used as fuel and the iron ore was placed right in the fire. To effectively remove the unwanted oxygen from the ore, the bloomery needed to have quite a lot of carbon monoxide. Because CO is a byproduct of combustion, in particular when the coal doesn't combust all the way, it was important to control the oxygen content in the bloomery. A bellows was used to make sure there was just enough oxygen to feed the fire and keep it hot while also ensuring that CO remained the dominant gas. In this environment, the needed redox reaction occurred and the unwanted parts of the ore, the slag, melted and dripped to the bottom of the bloomery. The temperature was hot enough to melt away the slag but not hot enough to melt the iron. The result of this process was the *bloom*, a chunk of iron chock-full of holes and impurities trapped within those holes. The iron was then hammered at a temperature around 1000°C (depending on which authoritative source you ask) to bang out as many of the impurities as possible, creating *wrought iron*, meaning iron that has been worked.

To make steel from iron, carbon needs to be added through a process called *carburization*. Charcoal has an extremely high carbon content, and steel is just iron mixed with carbon, so how come a bloomery doesn't end up producing steel instead of elemental iron? Unfortunately, it's not that simple. Refining steel is a highly temperature-dependent process. At lower temperatures, the carbon doesn't infiltrate the iron much, but as the temperature rises, more and more carbon is incorporated. You'd think this would continue until the melting point of iron was reached, at which point it would just mix together like copper and tin do for bronze. It turns out that alloys can be weird. Like, really weird. Many alloys have what's called a *eutectic point*. Before I explain this much further, I should say that part of my thesis work involved analyzing patterns formed by silicon–gold eutectics in the carbon nanowire growth process (more on that in the next chapter). In the experimental group that actually grew the eutectic patterns, only about half of them could say confidently that they understood what a eutectic was, and none of them could explain it to anyone. So, if you think eutectics don't make sense, you are in the company of more than a few people who go by "doctor." When two materials such as iron and carbon are mixed together, one would assume that the mixture wouldn't melt fully until the melting temperatures of both were surpassed; however, sometimes the mixture will melt at a point below either of the materials' melting points. This "easy melting" happens at a specific temperature and percentage composition. The liquid produced is always at that temperature and percentage composition. The point at which that occurs is called the eutectic point, and the liquid produced is called the *eutectic*, meaning "easy melt." In the case of steel, the iron absorbs carbon at roughly 1150°C until the composition is 3.5% carbon and 96.5% iron. At that point, it melts quickly. The steel produced is in liquid form and can be cast, but it is useless due to its high carbon content. This is *cast iron*. A bloomery can produce different types of iron, suitable for either a porch railing (wrought iron) or a frying pan (cast iron), but it can't produce steel. From here, there are several different methods used to turn iron into steel. I'll focus mainly on early Mediterranean steel production, given that Westeros is based more closely on that region than on China or India, the two steel powerhouses of the time. (If you are wondering how Valyria—or India and Syria, really—made such great steel, don't worry—there's a whole chapter on it.)

Southern Europe was great at a lot of things, but quality steel production was not one of them. The quality of steel is highly dependent on the carbon content. If there's too much carbon, the resulting alloy will break when hammered rather than being forged into useful swords; too little carbon and the steel isn't very strong, nor does it hold an edge well. It would seem that the carbon content could be controlled by regulating the temperature in the bloomery—and this is, in fact, the case—but back then, there was no way to accurately control the temperature. The problem of making steel from either wrought or cast iron can be attacked two ways. Either one must figure out how to remove carbon from cast iron, or one can try and force carbon to take up residence in wrought iron. Different regions chose different methods. China took cast iron and tried to pull the carbon out, while India took wrought iron and tried to shove carbon in. Southern Europe didn't really do either. Iron was smelted in a bloomery and then worked to remove the remaining molten slag and forged to form spongy iron. Forging is done at a temperature hot enough to allow any remaining carbon in the iron to be drawn out by surrounding oxygen, leaving nothing but iron. Once the bloom has been pounded into a solid mass, blacksmiths would add carbon back in to create steel, but they hadn't perfected a method to do it effectively. The iron was reheated in a reducing environment, meaning one with lots of carbon monoxide, for a long period of time. Carbon would mix with the iron on the surface of the sword and create steel; however, this was only about a millimeter's worth of surface steel. It was better than nothing, but it still wasn't a full steel blade.

In China, things developed much differently. The Chinese decided to take cast iron and remove the excess carbon. The Chinese were excellent at creating cast iron using a blast furnace. This type of furnace quickly forces air through a chamber filled with excess carbon. This quickly made cast iron by both having the requisite carbon readily available and quickly raising the temperature. From there, they heated the cast iron for an extended period of time. This caused the carbon to turn into graphite that set into either steel or iron. Next, the so-called blackheart was repeatedly heated, forged, quickly cooled, and folded over itself, and the process was repeated (sound familiar?). Each time it was forged, the air bonded with the outside layer of iron creating iron oxide. As this oxide was folded back into the bulk steel, it pulled the graphite out of the matrix and gradually pushed it out of the steel. Eventually, this created wrought iron. As with all other processes, it was hard to control the carbon content, so a process called *co-fusion* was used. Wrought iron was wrapped around sheets of cast iron and heated in the blast furnace. As everything melted, the extra carbon from the cast iron mixed with the wrought iron to create steel. The carbon content was also partially controlled by regulating the alloy's weight. The Chinese produced excellent steel by this method.[2]

FROM STEEL TO SWORDS

The production of steel continued to evolve throughout the Middle Ages. The Europeans further developed the process of carburizing wrought iron from the bloom. Blacksmiths found that by leaving the wrought iron or bloom in a charcoal fire for certain amounts of time at certain temperatures, they could control the amount of carbon that would flow into the iron. The longer it was left in and the hotter the temperature, the more carbon would be incorporated in the steel. The downside of this is that the carbon content wasn't uniform. Though blacksmiths of the day didn't quite understand why these techniques worked, modern scientists do: the crystal structure of steel below the melting point is highly dependent on temperature.

There are different crystal structures based on how hot the metal is. Steel is an interesting alloy in that it has both eutectic and *eutectoid* points. A eutectoid point is the temperature that causes one type of solid to break into two others. Steel, for example, takes a form called *austenite* at temperatures above 900°C, when it is still in solid form. Austenite has an FCC structure like copper. Just as copper takes in tin, austenite takes in carbon. It can take in much higher volumes of carbon, more than 2%. As the austenite cools, it goes through a eutectoid transition and breaks into ferrite, which is a fancy name for iron in a BCC structure. The carbon has to go somewhere, so it mixes with iron to become *cementite*, which has a carbon content of about 6%. As you can imagine, cementite is very brittle; in fact, it can be classified as a ceramic. The ferrite and cementite don't form islands the way tin and copper might. Instead, they form layers. This microstructure is called pearlite and is very strong. When worked, the additional grain boundaries can increase the strength of steel by more than four times.

The discovery of *quenching*, or quickly cooling the steel, further increased its hardness, leading to better, stronger swords. If the hot sword is quenched, it doesn't go through the eutectoid transition. Instead, it forms *martensite*, which is even harder than pearlite. The standard tradeoff applies, however—the material is much more brittle. Because the steel is being cooled so quickly, the carbon can't order itself with the iron to create cementite. Martensite is a BCC crystal under a whole lot of stress from the carbon atom that is replacing some iron in the crystal. As with tin atoms in bronze, this strain makes the alloy stronger. The quenching process takes less than a second to form martensite. With a temperature drop of that speed needed to create martensite, something the size of a sword could not cool that quickly in the core. For a blade, the outside would be hard yet brittle martensite with a core of less brittle pearlite. It's unclear when quenching came on the scene, but for those who have read Homer's *Odyssey* (or the *Percy Jackson* series), you know it must have been before then. Odysseus has a run-in with a cyclops named Polyphemus and "quenches" a burning tree in his eye. Confused yet? Don't worry—figure 5.2 shows a phase diagram that sums everything up in a graph. It's not confusing in the slightest and it certainly didn't take me hours to fully understand.

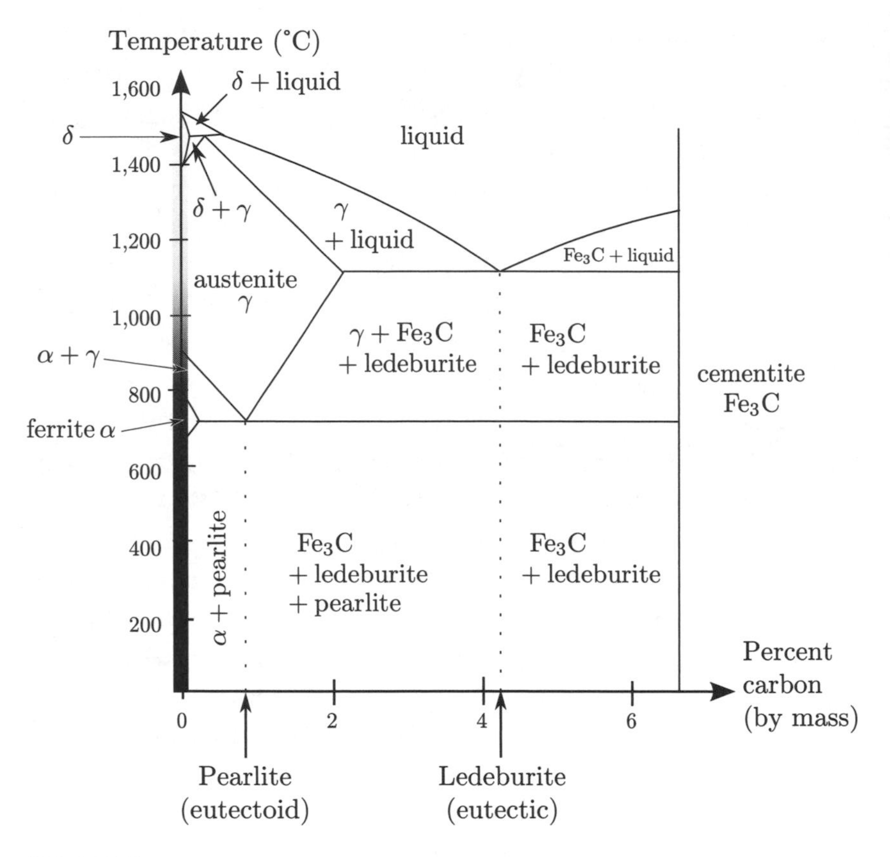

FIGURE 5.2
Phase diagram of steel. The *x*-axis is the percentage of carbon and the *y*-axis is temperature (°C). The phase diagram can be used to find the structure at a certain temperature and carbon content.

The martensite of quenched steel is still fairly brittle, which isn't super great for a sword. To counteract this, a blacksmith could *temper* the steel. He would heat the blade to a temperature below the eutectoid point so the martensite couldn't change back into austenite. As it was heated, the carbon atoms in the steel would move around. Generally, they would move toward each other and form islands, as seen in bronze. As with bronze, this made the steel tougher.[3] There's a sweet spot between hardness and toughness for holding an edge. If the edge is tough and strong it can bend upon striking. If the metal is very hard, it won't bend but it might break when hit. Tempered martensite seems to hit that balance well. Damascus steel did much better, so keep reading on to chapter 6. Through experimentation, blacksmiths were able to create steel with controlled properties. They may not have known the fancy terms for what they were doing, but they were able to create weapons (and tools) that were hard enough, strong enough, and tough enough for their purpose.

How Does It Do in the Cold?

Normally, those wielding swords would not have to ask this question unless fighting in Antarctica. But the inhabitants of Westeros are not that lucky. It's bitterly cold north of the Wall, and this will have an effect on normal steel. Alloys have a point called the *ductile-to-brittle transition* where, well, it behaves exactly as advertised. This is the threshold where the metal goes from being able to absorb impact to shattering instead. In the case of steel, the amount of energy needed to break it starts to decrease rapidly at temperatures below 0°C. Different types of steel have different rates of decline, but in general, steel doesn't do well in the cold. It's difficult to find peer-reviewed research on how ancient steel behaved in the cold because no one really cares, but it's important for modern steel for obvious reasons, so I'm basing this discussion on those results. Eventually, the energy needed to break steel drops sharply—this is the ductile-to-brittle transition temperature, which for modern steel occurs around −40°C. Some steel, however, loses half its strength by the time it reaches 0°C.[4] This temperature dependence is caused by the BCC crystal structure. It's much easier for an FCC crystal structure to slip than it is for a BCC structure. As BCC structures cool and the atoms move closer together, the internal stress also increases. This doesn't happen with FCC lattices. Because steel is a BCC structure, it becomes brittle when cooled significantly. This is really terrible news for those headed north of the Wall.

One of the most dramatic examples of this is the sinking of the RMS *Titanic*. In 1912, steel was strong and easy to acquire, but there was little quality control. Today, those sailing in frigid waters can be sure the steel has been engineered to deal with those conditions. The less carbon in the steel, the more extreme the temperatures it can handle. Back in the day, however, there were no such assurances from steel manufacturers. If the *Titanic* were in warmer waters, it would not have sunk. But because the steel was in ice water, its strength was greatly reduced. In warmer water the hull would not have been breached in the way it was; there would have been some massive dents, but not a break. The steel would have been tough and ductile enough to absorb the force of the iceberg without breaking. In the frigid temperatures it was sailing in, however, the steel was so cold it had become brittle enough to be shattered by ice. (On a different *Titanic* topic, if you are interested in the math behind one of the great mysteries of the movie—there was room on the raft!—hit me up on Twitter and blame buoyancy.)

SAM VERSUS A WHITE WALKER, TAKE ONE

A classic *Game of Thrones* scene that involves a sword is the cold open of the season 4 premiere, the aptly named "Two Swords," in which Ned Stark's sword Ice is cast (which, if you remember, means melted and poured into a mold) into Widow's Wail and Oathkeeper. Even though Ice was forged from Valyrian steel (which I'll talk about in the next chapter), bear with me here, because this is relevant. Watching House Stark's ancestral sword being broken, melted, and poured into two molds is a deeply dramatic moment. Molten, glowing metal slowly flows into the molds of swords meant to kill Starks. From a narrative perspective this is an amazing scene, seeing the Stark legacy destroyed by one of the three blacksmiths who can still work Valyrian steel. From what I've seen, though, he is the worst blacksmith in the Known World. If I've done my job in this chapter, you already know why. First, iron doesn't melt in a fire like that. Even coal fire can't get hot enough to melt iron. That's basically why it took thousands of years to produce effective steel. Second, stone will cool iron quickly, so it wouldn't flow as beautifully as it does on TV. Third, even if they did get the steel to melt, it would be in eutectic form, or cast iron, which is useless as a weapon. Brienne would have been killed by the Hound with one swing after dramatically watching her sword shatter. Finally, why on Earth would someone try to cast a three-dimensional object in a two-dimensional mold? That's just silly. Bronze swords were indeed cast, but ancient smiths used three-dimensional molds because they were smart. I say all this in this chapter rather than the next because it applies to regular steel, too. If this smith is representative of the crackerjack team of smiths in Westeros, no wonder Sam had a rough day with that White Walker. The real surprise is that any of their swords worked at all.

From here, there are two ways to approach the question of Sam's sword shattering at extremely cold temperatures when he hit the White Walker. If his sword was indeed cast in the way Oathkeeper and Widow's Wail were, it's not surprising it broke on impact with a Walker. What is surprising is that the sword made it long enough to fight a Walker in the first place. If it was cast, it's possible that the Chinese method of co-fusion might have been used; however, this still wouldn't account for the sword being cast directly from cast iron. The other way to look at it is by dismissing the rogue blacksmith from Volantis and assuming Sam's sword was made by carburization, which would result in a blade with an iron core and only a thin outer layer of steel. If his sword was indeed made via the more effective methods of quenching and tempering, it's still going to have a bad night with a Walker. As I described earlier in this chapter, the crystal structure of both iron and steel causes them to lose their durability and become brittle at low temperatures. Just ask Jack Dawson. Heck, if normal people in Iowa can accidentally rip a handle off a car door, Sam and his normal sword didn't stand a chance in the frozen North. Good thing he had dragonglass and Jon had Longclaw.

6

Valyrian Steel, Made in Damascus

The blade was Valyrian steel, spell-forged and dark as smoke. Nothing held an edge like Valyrian steel.

—A Game of Thrones

For the majority of the inhabitants of Westeros and medieval Europe, normal steel did just fine. It wasn't perfect, but it got the job done. Valyria, however, stepped up the steel game. They devised a method to make steel that was harder, stronger, lighter, and apparently great for dispatching White Walkers. Though in Valyria it was made with dragon fire and spells, which helped give it its superior quality, there is a real-life analogue to Valyrian steel. In the last chapter I talked about how southern Europe and China took iron from ore to blade, but I specifically left out India—that is, the real-life Valyria. India perfected a method of creating steel that could be fashioned into blades that were light and had an amazingly sharp edge and beautifully patterned blade. This steel was sought after by the whole world, but it was Damascus that became the place to turn this steel into blades. The working of the steel created the beautiful trademark ripple pattern on the blade, and it was said the edges were sharp enough to split a human hair in half lengthwise. Like Valyrian steel, these blades were a useful status symbol, passed down through generations of families. Also like Valyrian steel, the methods for making Damascus steel have been lost over the years. Scientists, like the smiths of Westeros and Essos, have not been able to recreate the metal, although they are still trying. There are competing theories as to how one might create modern Damascus steel. The remaining blades are extremely expensive, but unlike Valyrian steel, you can buy knockoff Damascus steel on eBay.

After learning all about the steelmaking process in the last chapter and knowing the reasons why it is so complicated, what set the steelmaking in India apart? The best steel is made using a method that accurately controls how much carbon is in the steel and what atomic form—pearlite, martensite, cementite, or ferrite—the final product takes. This could be done in the steelmaking methods I talked about in chapter 5, but not with precision. The smiths of India, however, developed another method that introduced carbon into iron in a more controlled way. This so-called *crucible steel* had a better composition than other types of steel. They then figured out how to use heating, cooling, working, and quenching to produce the best combination of cementite, pearlite, austenite, and martensite. In addition, the smiths of Damascus and India accidentally created some of the most advanced materials science of the day. There are two groups of scientists that are in a bit of a tiff about what really made Damascus steel so special. In 1980, two materials scientists from Stanford University, Jeffrey Wadsworth and Oleg Sherby, developed the eponymous Wadsworth–Sherby recipe for making modern-day Damascus steel. J. D. Verhoeven from Iowa State University worked with a number of blacksmiths, mainly Alfred Pendray, to develop another method in 1998. The two groups went through a number of publications and rebuttals, and I can't say the issue has been definitively solved. Taken together, the two groups' work, as well as a more recent study by a group in Germany, gives a pretty complete picture of the process and final product.[1] This argument took up hundreds of pages in scholarly publications, and only seemed to end when the combatants either died or retired. I wish I could go into the psychology of academic jousting, but alas, I have neither the time nor the training. To be clear, Valyrian steel, like real-life Damascus steel and regular steel, must be forged, not cast. At least in the books, these terms are used correctly: Ice is described as being "reforged." From here on out, I will assume that no one on the production team of the show bothered to look up the difference. Let's all forget that dramatic yet incorrect scene.

RAW MATERIALS: CRUCIBLE STEEL AND WOOTZ

Damascus steel got its name much the way Greek fire did—as in, there's no real reason it should be named Damascus steel because all anyone in

Damascus ever did was sell the stuff. The raw steel cake, called an *ingot*, was created in India, forged into blades in Persia, and eventually sold in Damascus. Europeans first learned of the steel during the Crusades, when they were on the pointy end of these superior weapons. The process for making this type of steel and then being able to work it into blades wasn't really figured out by the Western world until around 1982. Although blacksmiths of Western Europe purchased ingots, they were not able to work the steel into weapons due to its brittle nature. The best way to understand what made Damascus steel so special is to follow it from iron ore to finished blade. Unlike the bloomeries discussed in chapter 5, the precursor to Damascus steel blades was *wootz steel*, a type of crucible steel.[2]

The first step in creating any steel is smelting iron. In the West, this was done in bloomeries that could get hot enough to separate slag from elemental iron. They were not hot enough to melt the iron, however, unless the iron was at its eutectic point, at which point it would become cast iron. Recall that, in this method, the ore was in the same chamber as the fire, and that carbon and CO were involved in the process. By contrast, the crucible steel process starts out the same way as bloomery iron, with iron ore being smelted into bloom. The bloom was then placed in a *crucible*, or small clay pot about 8 inches high and 2 inches in diameter with a 0.25-inch-thick wall. Wood chips, specifically *Cassia auriculata*, and a fresh leaf or two from the *Convolvulus laurifolius* plant were also added. Later, it was determined that Sorel, an iron–carbon alloy that contained trace amounts of other elements important to the process, was also added to the crucible. These tiny details are important. I also like to bake and make candy, and I can tell you that recipes count and tiny differences in temperature and ingredients can make the difference between an exquisite end product and glop.

The crucible was then placed in a fire and the iron inside was heated to its melting point. Yes, I did indeed just spend a chapter telling you that it wasn't possible to get a fire hot enough to melt iron. This was only sort of true. It's not possible to get a *useful* fire hot enough to melt iron in a bloomery. That's a lot of qualifiers. Remember that the environment in a bloomery had to be high in CO and low in CO_2. High CO concentrations pull the oxygen out of the iron or reduce it. Once the ratio of CO to CO_2

falls below 3:1, the reaction changes from reducing iron to oxidizing, at which point it's the exact opposite of what they wanted to happen. The problem is that CO is created through incomplete combustion, meaning the fire isn't burning completely and is not as hot as it could be. For a fire to reach 1400°C, it needs to be burning in an environment with a CO to CO_2 ratio of about 2:1. In a bloomery, the fire and the steel are in the same chamber, meaning that the combustion environment of the fire is the same environment as the iron. The heat of the fire has to be kept in a range that won't create an environment that will oxidize the iron. When making crucible steel, however, the environment in which the iron is being heated is completely independent of the fire. The ratio of CO to CO_2 can be whatever is needed to achieve the desired temperature. Modern-day smiths have been able to use ancient methods to heat charcoal fires to temperatures above 1500°C.

The crucible was placed in a charcoal fire and then covered with coal. There are different accounts of how long it would be left there, but it ranges from 2.5 hours to 24 hours. I talked about the eutectic point in chapter 5 and mentioned that at approximately 1100°C the iron takes in carbon until it reaches a concentration of 4% and melts out. This wasn't a complete explanation, but it was all that needed to be said for that chapter. Look at the boundary line between liquid steel and a liquid–austenite mix. If the temperature is heated right to that line and the steel is melted, it will take in the corresponding percentage of carbon. In the case of 2% carbon, that hits at right about 1400°C. They probably didn't have this super confusing phase diagram back then—they just knew it worked. Damascus steel has a very high carbon content (roughly 1.5%–2%). Scientists and historians know for sure that the iron inside the crucibles was melted. If the fires couldn't go beyond 1300°C, the liquid produced would be the eutectic, with a carbon content of about 4%. From there, carbon would have to be removed in a secondary process to get the concentration down to 1.5%–2%. If you look back at the phase diagram of steel (figure 5.2), you'll see that at a concentration of 2% carbon, steel needs to reach a temperature of about 1400°C to melt. Initially, historians thought there must be a process to remove carbon—decarburization—involved because fires couldn't get hot enough to melt 2% carbon steel, but there was no real historical evidence for a secondary process. Given the steel being in a crucible and a coal fire's potential high temperatures, it was possible—and, indeed, likely—that the liquid in the crucible was steel with a carbon content of roughly 2%. After waiting 2.5–24 hours and checking to make sure the contents were liquefied, the crucible was removed from the fire and left to cool naturally; in some cases, water was thrown on them. The end product, a solidified mass the size of a hockey puck, was the ingot. It was often described as having a crystalline pattern on the top.[3]

Originally, historians thought it was a pure mixture of iron and carbon, but recent developments in materials science and imaging techniques have debunked that theory. The addition of Sorel, leaves, and wood chips, as well as the ore that was originally used, infused the steel with trace amounts of 14 other elements. Three different groups measured the concentration of these additional elements and came to roughly the same conclusions: the trace elements consisted of 0.15% phosphorus, 0.03% manganese, 0.06% sulfur, 0.06% silicon, and smaller amounts of nickel, cobalt, chromium, copper, molybdenum, tungsten, niobium, aluminum, and, most importantly, vanadium. In their attempts to recreate Damascus steel, Verhoeven and Pendray were able to create a similar-looking steel on only a small number of occasions. The majority of the time, the blades were missing the characteristic wavy pattern (figure 6.1). Verhoeven and Pendray realized they hadn't been using Sorel iron, even though it had been listed in historical records as an ingredient in the ingot. In their paper, they admit this was a gross oversight. When they added the Sorel iron to the ingot, they were able to create blades with the characteristic pattern at a much higher rate of success. They wanted to find out what was so special about Sorel iron. After determining the composition of the final steel, they sought to identify which of the 14 trace elements was needed to create the pattern. Vanadium was found in both the Sorel iron and the ingot, but because it was found at such low levels, Verhoeven and Pendray ignored it for several years. Eventually, they started the process with normal iron and slowly added in the trace elements until they found the ones that best resembled the patterns on true Damascus steel. They focused mostly on the *carbide-forming* elements, meaning the ones that will bond with carbon. Of the trace elements, they tried vanadium, molybdenum, chromium, niobium, and manganese, with vanadium

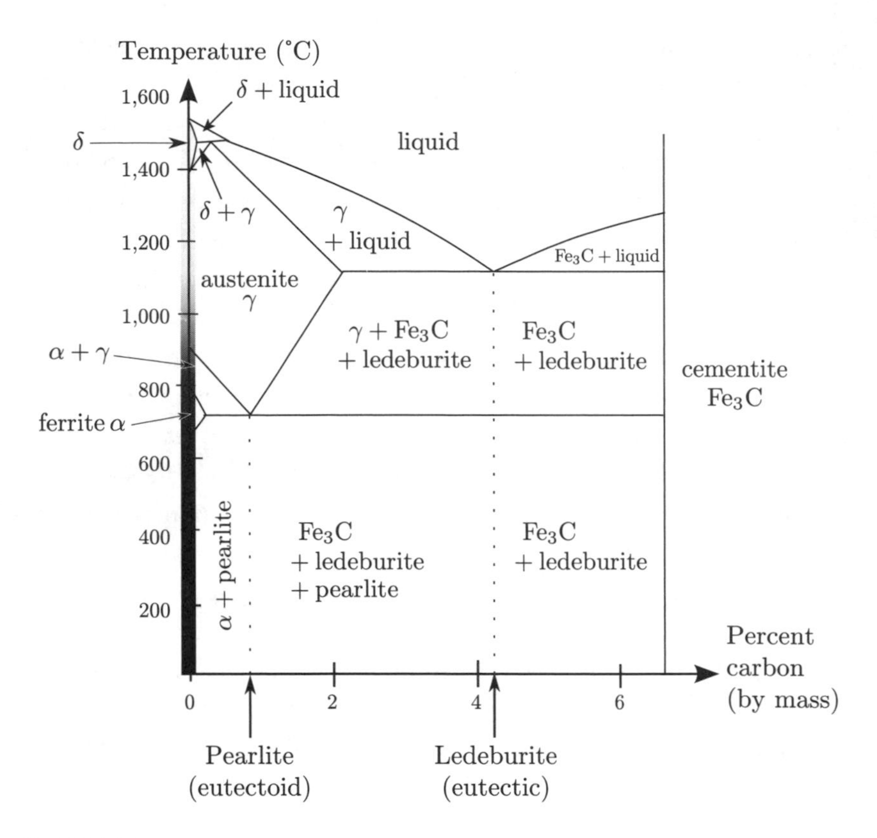

FIGURE 6.1

Phase diagram of steel. Wootz steel, the precursor to a Damascus steel blade, has a carbon content of 1.5%–2%. α is ferrite (or iron), γ is austenite, and Fe_3C is cementite. Martensite is not seen on the diagram because it is formed only through quenching.

and molybdenum being the most important. In particular, the vanadium helps the cementite particles orient themselves in a single direction instead of all willy-nilly. It's this order that creates the distinctive rippling pattern. I'll talk more about how cementite and other structures affect the properties of the blade in just a minute.[4]

WORKING INGOT: NOT EUROPE'S SHARPEST MOMENT

Though many were able to purchase wootz ingots, very few were skilled enough to actually do anything with them. The Europeans were said to have had their finest blacksmiths attempt to forge (or take from ingot to finished blade) the famous swords to no avail. In fact, as far as I can tell from my research, it wasn't until 1982 that anyone in the West was able to do it.[5] The barrier was the fact that 2% carbon was just too brittle for western blacksmiths to work with. They were used to working with steel with a much lower carbon content, which made it more ductile and therefore less hard and brittle. When they tried their normal techniques on wootz steel, it simply shattered under their hammer. This issue and any potential solutions come down to the phase diagram for steel. This was included in the previous chapter, but I'm including it again here for easy reference (figure 6.1).

In Europe, they worked with steel at approximately 1300°C. If you remember from the last chapter, this was the range in which steel was in its austenite phase. When the blade is cooled from there, it drops down into the pearlite region, or it can be quenched to form an outer layer of martensite with a pearlite core. Now, look at that same temperature but move over on the *x* axis to the 2% carbon label. You'll see that at 1300°C the steel would be a combination of austenite and liquid steel. When it's hit with a hammer it shatters. You can see from the phase diagram that it would be better to work this type of steel in the austenite (abbreviated γ) and cementite (seen here as Fe_3C, its chemical formula) range of 700°C–900°C.

In a 1999 paper, Oleg Sherby discussed in detail what he called the Wadsworth–Sherby method for making Damascus steel blades. Verhoeven gave the method that name in a paper saying how it was wrong, just to make sure the world knew exactly who was wrong.[6] Verhoeven argued that although the method produced blades that on a microscopic level looked and behaved like Damascus steel, imaging techniques showed that the structures were different at the microscopic level. Wadsworth and Sherby then rebutted his assertion by saying Verhoeven didn't know what he was doing when he followed the Wadsworth–Sherby method. In a 2001 paper, Verhoeven again asserted that the Wadsworth–Sherby method didn't work and described his own method and the role of impurities. What is interesting about Wadsworth and Sherby's method is that they don't mention how impurities might give rise to the banding seen. Verhoeven gave a pretty convincing argument that the impurities were needed to achieve banding, and this was not taken into account by Wadsworth and Sherby. What they both agree on, though, is that the blade, at 2% carbon, must be forged in the region where it is austenite and cementite, between 700°C and 900°C. But enough about the fight—let's hear about their methods.

Wadsworth and Sherby were the first to publish results saying they had achieved Damascus steel. In a 1999 paper, they laid out a three-step process for producing a blade.[7] First, the wootz is heated to about 1100°C, putting it in the austenite phase. It is then rolled to mix everything around and fully integrate the carbon. The wootz is also cooling down during this process. Second, the wootz is then reheated to 1100°C and left there for 48 hours. Being kept at this temperature for so long allows the austenite to arrange itself in bigger crystals and create longer-range structures. This means there are fewer but longer grain boundaries instead of lots of little ones. The steel is then cooled very slowly into the region of austenite and cementite. The austenite is already sitting in nice big crystals, and the newly forming cementite doesn't want to get in the way of that. Instead, it grows in long strips along the grain boundaries. At this point, the steel has been rolled but not really worked; if it was etched, however, it would be possible to see the characteristic stripes. Third, the wootz is then heated right to the boundary between the pearlite and austenite–cementite regions. This causes the cementite that didn't make it into the long strands to dissolve, leaving austenite and long networks of cementite. The steel is then rolled to break up these long networks. It remains ordered in strands, to some degree, but they are not as long. When cooled to room temperature, the cementite strands lay in similar directions and appear bright after etching on a background of dark ferrite. With this new structure, the blade has amazing strength and durability.

The method extolled by Verhoeven and Pendray is similar in many ways but different enough to cause a feud rivaling that between the Starks and the Lannisters. Both teams realized that working in the 700°C–900°C range was critical, but the biggest difference lay in their explanations of where the bands came from. Verhoeven and Pendray skipped step 2 of the Wadsworth–Sherby method and went straight to forging at the same crucial temperature. As they worked the steel, it cooled. Before it fell below the needed temperature, they heated it again and worked it again.

It took about 50 cycles of heating, working, and cooling to hammer the steel into a blade. They wanted to make sure to take into account the role of impurities, particularly vanadium, in their theory. They suggested that as the metal was heated, some cementite would dissolve, as theorized by Wadsworth and Sherby, but Verhoeven contested their hypothesis, instead suggesting that the formation of cementite strips was nudged along by impurities. These impurities prevented the cementite from fully dissolving in the liquid metal, allowing large particles to remain. Each time they underwent this process, the bigger cementite particles attracted more cementite and grew slowly. After about 20 cycles of this, it became possible to see the characteristic bands. In the end, both groups produced steel with cementite wires in a matrix of ferrite, causing visible bands. Wadsworth and Sherby make no mention of the role of impurities, while Verhoeven gives impurities credit as the critical ingredient for making Damascus steel. Wadsworth and Sherby give wootz with a carbon content of 1.8% as their starting point. They do not, however, say whether they are using wootz that they made in their lab containing *only* carbon and iron, or whether they obtained wootz that included defects, as historical wootz would.[8] Answering this question is the key to figuring out which group's theory is correct. In the end, they both end up with long bands of cementite, which turned out to be cementite *nanowires*, and a really cool-looking sword (figure 6.2). In 2004, a group from Germany decided to make a better image of the edge of a Damascus blade and see if everyone had been missing something. Turns out they had been.

Really Ancient Technology Meets Really New Science

The work by Verhoeven and Pendray and by Wadsworth and Sherby was carried out in the 1980s and '90s. Materials science has advanced significantly since then. The first paper in that series was published in 1982, the year the Commodore 64 was released. (That was a computer, not a band.) Since then, processors have gotten faster, cell phones are now powerful pocket computers, and the SUV was invented. The discovery relevant to this discussion, however, is the carbon nanotube. Yes, in just two chapters we have gone from the Stone Age to the Information Age. In 1991, Sumio Iijima of Japan was the first to discover carbon nanotubes. As with many

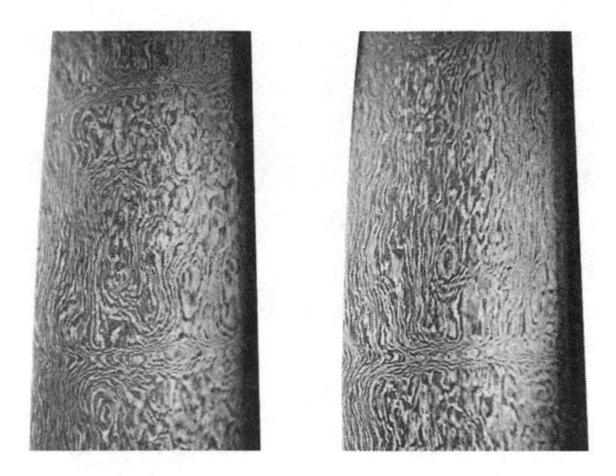

FIGURE 6.2
An 18th-century Damascus steel blade. The beautiful and characteristic swirl pattern is clearly visible.

major scientific advances, there is much controversy regarding whether or not he was the first to discover them, but he was indeed the one to make the scientific community take notice. Carbon is the most abundant element on Earth and it can be structured in many different ways. Graphite, diamond, and amorphous carbon are just a few of forms it can take. The properties of these different forms vary wildly. For instance, graphite is very conductive. You can draw with a pencil on a piece of paper and use it as a wire, albeit kind of a crappy one, but who's ever heard of conducting electricity through the diamond in your ring? Carbon also has the ability to organize into a flat sheet only one atom thick. This is called *graphene*, and its discovery won Andre Geim and Konstantin Novoselov the 2010 Nobel Prize in Physics. We've talked about BCC and FCC crystal structures, but graphene has a honeycomb lattice shape (like chicken wire) and a whole lot of interesting properties. When this single layer of carbon atoms is rolled into a ball, it becomes the famous soccer ball–shaped buckminsterfullerene (or buckyball), but if it's rolled into a tube, it's a carbon nanotube (CNT).

Carbon nanotubes have some pretty amazing properties. One of the pioneers of carbon nanotubes, and my personal hero, was Mildred Dresselhaus, affectionately known as "the Queen of Carbon." She specifically looked at the electrical properties of carbon nanotubes. (This isn't particularly relevant to a discussion of swords forged before the discovery of electricity, but she's awesome and more people should know about her work. You might recognize her from a wonderful GE commercial.) What we're really interested in with respect to carbon nanotubes is their strength and how they might form in steel. CNTs are actually much stronger than steel, even Damascus steel, and they are not only strong but also very springy. This is pretty much an ideal combo for weapons-grade steel. It takes five times more energy to bend a CNT than it does steel. Unlike steel, the strength comes not from the tiny dislocations and grain size but from the fact that it doesn't have any dislocations. A nanotube is a single crystal, and its honeycomb crystal structure isn't conducive to the slipping seen in BCC and FCC structures. The covalent bonds between the carbon atoms are extremely strong, which leads to a tensile strength that

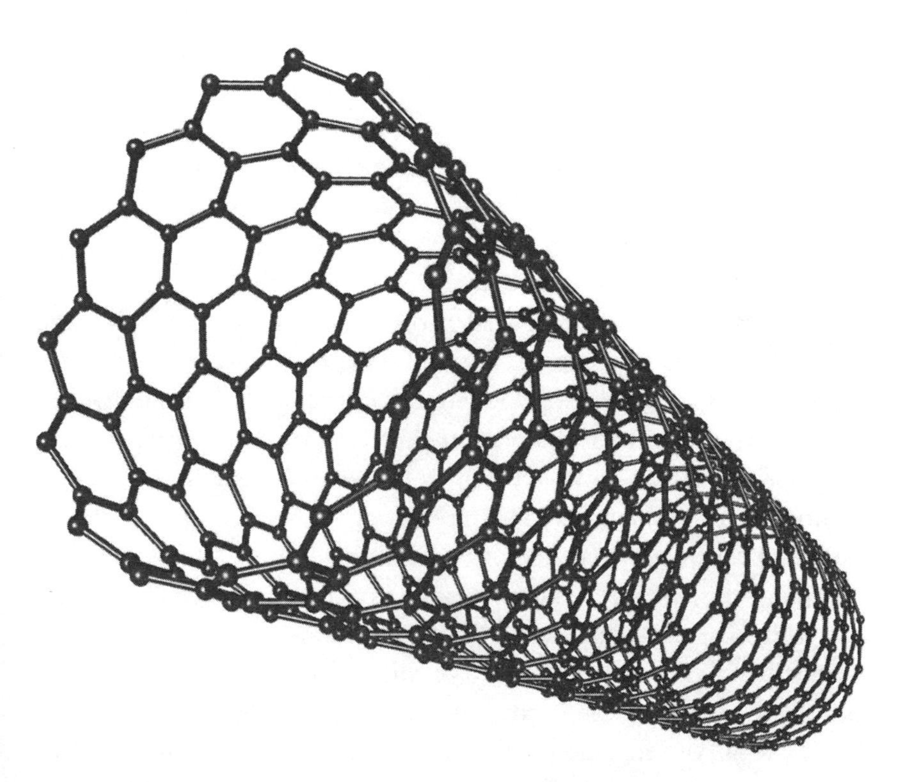

FIGURE 6.3
A carbon nanotube

is 100 times that of steel. If you do happen to bend one, it will bounce right back into place. These little guys are practically indestructible. And, as it turns out, they are found on the edge of a Damascus steel blade, reinforced with cementite.

Alexander Levin and his group in Dresden, Germany, began examining the edges of Damascus steel blades just as nanotubes and nanotube imaging began in full swing.[9] The team was not satisfied with the current explanations for the blade's strength and the emergence of its pattern. They chose to look at how the various elements were distributed throughout a blade to in hopes of getting a better understanding about the microstructure of the steel. They convinced a nice person at the Berne Historical Museum to give them a small piece of an actual Damascus steel sabre. This is the same weapon Verhoeven looked at. Since the piece was cut, they were able to image it both from the front and through a cross section.

They started out, like Verhoeven, by identifying which elements besides iron were actually in the blade. Levin's team ended up with numbers very similar to Verhoeven's but different enough to make them conclude the composition of the blade was not uniform. They looked at the crystal structure using X-ray diffraction, which involves bouncing X-rays off of a structure and seeing how they are reflected. This can give an idea of how the object reflecting the X-rays looks. In the blade, they found ferrite, cementite, martensite, and graphite. Though the martensite was distributed rather evenly throughout the blade, the cementite and the ferrite were found in different concentrations throughout their sample. Cementite concentration was higher on the outside of the blade toward the back. There wasn't much in the bulk steel, nor near the cutting edge. They also found that cementite was present in three phases: first, as part of the layered pearlite mentioned in chapter 5, and second, as grains of cementite, as expected, but also in cementite nanowires. This was the first time nanowires of cementite had been observed. Using transmission electron microscopy, a technique where a beam of electrons is shot through a sample and an image is recorded on the other side, they were able to look more closely at the newly discovered nanowires. The researchers found that all of the nanowires ran parallel to each other, but that they do not appear in all sections of the blade. They may line up locally, but they don't have a preferred direction in the bulk. This means that the nanowires may be aligned one way in a small section of blade, but each section on the blade could be facing in a different direction. In addition, these nanowires didn't appear in all sections of the blade. As seen in normal cementite, there is a way ferrite, cementite, and martensite crystals like to line up. This supports the theory posited by Wadsworth and Sherby. Levin's team theorized that the nanowires were formed by a method similar to that described by Wadsworth and Sherby, but that impurities served as nucleation sites, or places where tubes would start forming. This hypothesis incorporates the findings of Verhoeven as well.

A year later, the group took even higher resolution images of the blade edge.[10] They found that not only did the blade have cementite nanowires, it also had carbon nanotubes. They found these after dunking the blade in hydrochloric acid in a controlled way. This caused the cementite to dissolve, thus exposing the carbon nanotubes; however, they found that this happened only in places where the cementite hadn't fully dissolved. They concluded that the carbon nanotubes were protecting the cementite by encasing the nanowires. Another group of scientists showed that cementite could indeed crystallize inside a carbon nanowire.[11] It just so happens that molybdenum, a key impurity found by Verhoeven, has the ability to stimulate carbon nanotube growth.[12] This might bring the whole argument full circle. Thanks to Levin's and Lei Ni's teams, we now know that Wadsworth and Sherby and Verhoeven and Pendray were both on the right track, but neither had the full picture This is why science is so cool. The forging of the steel potentially pushed the necessary impurities around along the plane in which the steel was being worked. As the impurities moved in the same direction around different parts of the blade, they were able to nucleate more nanotube formation. It's possible that cementite was then able to crystallize inside the carbon nanotube. As soon as I learned about carbon nanotubes, I wanted to build a faster bike out of one, but I think I'd be happy to settle for a sword.

"DAMASCUS" STEEL ON EBAY

By this point, you might have tried to find out where you can get one of these awesome blades for yourself. You might have entered "Damascus steel" into Google and got more than a few listings for things made of "Damascus" steel, everything from hunting knives to wedding rings. You also might think I'm full of crap for going on and on about how special this stuff is, when for a few hundred dollars you, too, could own some real-life Valyrian steel. Here's the catch: the internet stuff isn't real Damascus steel. It has the telltale pattern and sure looks the same. It's even called the same thing. But, unfortunately, it's not made the same way. It's a completely different kind of steel called welded or pattern steel. The goal is to make something pretty that is also functional, not make the world's most kick-ass sword that also happens to look cool.

Perhaps I am not being totally fair. The first pattern welding was done with very early steel to even out the problems of inconsistency. In the last chapter, I talked about how it was not easy early on to make steel and those who did make it couldn't usually make good steel consistently because it was hard to control both the temperature of the fire and the carbon content. The Chinese tried to overcome this problem by wrapping wrought iron in cast iron and letting them mix. It was a pretty successful practice. Pattern welding was even simpler than this co-fusion system. Blacksmiths would take multiple sheets of steel, each with slightly different properties, and heat them all up on top of each other. Then, they would be forged together into a single blade. The idea was that if one layer was hard and another was tough and they were forged into a single blade, the sword would have macroscopic properties between the two. The Celts were particularly fond of this method. Different layers had slightly different colors of steel since the final composition of the steel varied depending on how the bloomery was operated. As the steel was forged together, the different colors of steel wove together to create a beautiful pattern. The blade may then be etched or ground to further bring out the pattern. Smiths began twisting metals together instead of laying them flat to create even more intricate patterns. Eventually, as steel production improved, the creation of patterned steel weapons was a case of form over function. Seeing that it is even used in jewelry nowadays, modern pattern-welded Damascus steel is a decorative item rather than one that is prized for its cutting ability. I think the current pattern-welded steel is beautiful and a nice addition to any home or sword collection; however, one should be aware that it won't cut through a falling human hair, nor will it kill a White Walker.

Valyrian Steel and White Walkers

Valyrian steel and Damascus steel seem very similar. Both are made from steel that is unique to a region, and both are both are highly valued because of their unusual properties and difficulty to forge. The method for creating a Valyrian steel ingot (assuming it started as an ingot) has been lost, but the crucible method of forging Damascus steel was well known. In both cases, there are few people who know the specific recipe to forge a blade. Both types of steel are known for their distinctive swirl pattern and both are worked a number of times at specific temperatures to produce the precious final product. I think it has been assumed, if not outright stated, that dragons were involved in the production of Valyrian steel, either in the original smelting or crucible process or in the actual forging. Considering Westerosi smiths were still able to work with the remaining metal, it would make sense that the dragon fire was involved in the front end of the process and not the forging. It is unclear whether dragons can fully control the temperature of their flames. Up to this point, they'd just want it as hot as possible, but if they could control the temperature,

Figure 6.4
Modern pattern-welded blade from Toledo, Spain

they could produce steel with well-regulated carbon content and crystal structures. Seeing as this was a major hurdle in producing real-world steel, dragon thermostats would really help. I think by now we can assume dragons give materials a healthy amount of magic with their fire, which can't hurt either. It is also said that the blades were quenched in blood, in line with House Targaryen's motto. This isn't far from reality, either; some Persian texts recommended quenching the blade in the belly of a slave. There were also texts saying it should be quenched in the urine of a red-headed boy, so maybe wildlings *would* have a job if they moved south of the wall.[13] In general, and despite the dragon fire, it's really fascinating that GRRM's "fictional, magic metal" isn't all that fictional after all.

The biggest question I had about Valyrian steel, or Damascus steel, is how it would it behave at low temperatures, specifically those in which regular steel fails against White Walkers. As you can imagine, there's no research into how temperature affects the strength of Damascus steel because it doesn't get that cold in the Middle East. No one is going to build a ship out of it. There is probably a ductile-to-brittle transition, as with normal steel, given that most of the blade is made of a similar material. But carbon nanotubes, which protect and strengthen the cementite and allow the blade to hold an edge, becomes stronger at lower temperatures.[14] Also, CNTs do not have a ductile-to-brittle transition because they are formed with single-layer honeycomb lattice rather than BCCs and thus don't have the weaknesses brought on by dislocations. These nanotubes make up only a very small percentage of the Damascus steel blade, so the addition might not do much. However, seeing as they are protecting the cementite nanowires, they might be able to hold the crystal together, similar to how chicken wire is used to reinforce concrete. I haven't been able to find research on this specific question, but I, for one, am going to hope it's the case. I'd be happy if it was spells *and* science holding this one together, unlike the Wall. We'll never officially know if Damascus steel could kill a White Walker, but I'm pretty glad that remains a mystery.

7

Dragon Biology

Bats, but with Fire

Above them all the dragon turned, dark against the sun. His scales were black, his eyes and horns and spinal plates blood red. Ever the largest of her three, in the wild Drogon had grown larger still. His wings stretched twenty feet from tip to tip, black as jet. He flapped them once as he swept back above the sands, and the sound was like a clap of thunder.

—Daenerys Targaryen watching Drogon, *A Dance with Dragons*

Before I get too deep into the discussion of whether or not Daenerys Targaryen could fly away from the fighting pits on Drogon's back, I need to resolve an issue raised by many an Internet forum: Neither Drogon, nor his parents or siblings, are dragons—or at least not in the traditional sense. In traditional lore, dragons are fire-breathing four-legged creatures, whereas Dany's dragons only have two legs. Adding in another set of legs would add another set of joints and a whole new set of complications; not to mention that there are no four-legged flying creatures in nature. Targaryen dragons are a cross between wyverns and dragons. Wyverns have two legs, batlike wings, and a venomous bite, and are often associated with cold weather. They do not, however, have the ability to breathe fire. George R. R. Martin, D. B. Weiss, and David Benioff have always referred to their creatures as dragons, so please do not pillory me on Twitter for doing the same. Besides, "Mother of Wyverns" doesn't sound nearly as formidable.

Regardless of whether they are dragons or wyverns, they are often depicted as winged lizards. This brings up another tricky question: Are dragons warm-blooded or cold-blooded? With the exception of the tiny tegu, who only turns on the heat occasionally, lizards are cold-blooded. They have no ability to regulate their body temperature and must rely on the environment to keep their body temperature stable. This is why reptiles sun themselves during the day and are sluggish at night. Like most creatures, they don't function well when they are cold because, in their case, their metabolism slows way down. And, unlike humans, they don't have the ability to turn up the thermostat.

WARM-BLOODED OR COLD-BLOODED?

By all accounts, dragons should be cold-blooded. They have scales instead of fur or feathers and they lay eggs, typical hallmarks of a cold-blooded animal. Tegu excepted, only two warm-blooded creatures have scales: the pangolin and the armadillo. Both are adorable and very unlike dragons; also, neither is larger than a puppy. The largest flying creatures of all time, the pterosaurs, were cold-blooded. Cold-blooded animals convert much more of the energy from the food they eat into body mass. They aren't using that energy to heat and cool their body so they have a lot more left to use to bulk up. Dragons are huge so it would make sense that lots of the energy from the food they eat is transformed into body mass. It would seem this is an open-and-shut case: dragons are cold-blooded.

Well, not if you want a story of ice and fire. A story of just fire, sure, but introduce ice and there's a problem. In hot environments, cold-blooded dragons would be unstoppable; however, they have a huge problem once they hit ice. Muscle movement is controlled by chemical reactions in the muscles, and these chemical reactions happen much more quickly when the body is warm.[1] To move muscles and sustain life, enzymes in cells catalyze chemical reactions to break down glucose and proteins to provide energy for the cells to function. Enzymes are a bit picky as to when they like to catalyze these reactions. If the cells are too hot or too cold, the reactions slow down and the animal is sluggish. For cold-blooded animals, this means that in cold weather they have no ability to use the energy in their food to move, so they usually hibernate until they can warm up their bodies and get the enzymes to do their jobs.

When cold-blooded animals hibernate, they do it very differently from warm-blooded animals. Warm-blooded animals curl up in a safe space and live off their body fat until it's warm enough out for them to regulate their body temperature well. They can lose up to 30% of their body mass during hibernation. Hibernation in cold-blooded animals operates very differently on the cellular level. First, they do everything in their power to stay as warm as possible, such as burying themselves in mud and slowing their metabolism and heart rate so dramatically that they appear to be dead. They are very much alive, just slightly frozen on the inside. Animals can be approximated as big bags of water, so freezing solid in cold weather is a bit of an issue. However, some cold-blooded animals have found a unique way around this hurdle by letting some of the water in their bodies freeze so the rest can stay in liquid form. During hibernation, the water around the animal's cells freezes. Nature doesn't like two solutions with different concentrations of dissolved particles to be separated by a porous or permeable membrane, such as a cell membrane. If this happens, water moves across the membrane to even out the concentrations. This process is called osmosis. When the water outside cells freezes, it raises the concentration of the particles dissolved in liquid water and water rushes out of the cells to fix the problem. As water rushes out, molecules such as glycerol and urea rush into the cell. Just like salt on icy roads, this changes the freezing temperature of the water inside the cells and stops it from freezing, which would cause the cells to burst because frozen water expands. Reptiles create their own antifreeze.

Now, back to dragons. If dragons were cold-blooded reptiles, they would have to hibernate when they entered the cold climes of the North. They would appear to be dead in such frigid temperatures, and they certainly wouldn't be able to fly. Instead of fighting White Walkers, they would have to spend their time sunning themselves in Dorne. For any kind of story of ice and fire to work, dragons must be warm-blooded because they need to function—and function quite fiercely—in the coldest of conditions. (Not to mention that whole fire-breathing thing.) Seeing as they have scales and not feathers, we can assume they are a third type of scaled mammal, much larger and much less cute than the tiny pangolin. There are also two mammals that lay eggs, the platypus and the echidna, so there is precedent for that, too. In spite of their continued comparison to lizards, dragons are probably closer to the armadillo—a fearsome, flying, fire-breathing armadillo. With fangs.

Now that we've established that dragons can fight in the cold of the north, let's talk about how they'd get there. Could dragons really fly? And if so, how? Are they closer to an albatross, a fruit bat, or the long-extinct pterosaurs? Before we can make any comparisons, however, we need to know how things fly.

AIRPLANE FLIGHT

Though dragons have very little in common with a Boeing 747 other than size, I'd like to take a minute to explain how airplanes fly. One of the biggest misconceptions in physics is how aviation works, and I feel it is a public service to correct that misunderstanding in as many places as possible.

The classic explanation for how planes fly involves something called Bernoulli's principle, which says that faster-moving air has lower pressure than slower-moving air. Bernoulli's principle is important—just not in the way it's normally used. The common explanation is that a wing is shaped so that air has to flow more quickly over the top of it than the bottom, and the difference in pressure creates lift. There are two problems here: one, to create enough lift to fly an average-sized commercial jet, the wing would have to be absurdly rounded on top; and two, being the niece of a former professional barnstormer, I can say with great certainty that planes can fly upside down. I won't go into any more detail here because googling "Why do airplanes fly?" will give you a number of hits that will correctly (and incorrectly) explain plane flight quite effectively—just ignore any that involve Bernoulli's principle.

So how do planes *really* fly? It's fairly complicated, but there are two major components to flight. The first is that wings force air downwards. Newton taught us that forces come in pairs; if you want something to go in one direction, whatever's pushing it has to feel a force in the other direction. In the case of flight, this means that if you want to create upward lift, you have to push air downward. For something to leave the ground, it needs to feel an upward force that's greater than its weight. In the case of a 747, that's roughly 565,000 pounds of lift force. To create that force upward, air has to be pushed downward and fast. Force is equal to the change in momentum, so the downward momentum of air has to change in order to create lift. Momentum is equal to the mass (how much air is moved) times the speed (how fast you move it). Airplane wings are designed to force air downward and the faster they move, the faster they push the air and push themselves upward.

Air weighs a lot more than you might think: about 1.22 kg/m^3. Even weighing that much, a lot of air needs to be forced downward for a plane to fly. How can a wing divert that much air? An important detail to note is that airplane wings have an angle of attack. They don't fly head on—they are always tilted. The tilt allows the underside of the wing to push air downward, but the air flowing over the top also has a downward component. How is it possible to get air over the top headed down? It's done through the Coandă effect. When a fluid, such as water or air, flows into something curved, such as an airplane wing, it sticks a bit and flows around the curved surface instead of right by it. You can try this in your own kitchen by holding the back of spoon under the faucet and watching how the water curves around it. An airplane wing is slightly curved so air flows over it and down, and this downward motion pushes the wing up. The air that flows over and down from the wing is called the boundary layer. It's tricky to get the angle of attack just right. If it's too low, the plane won't get enough lift, but it's too high, the wing's tilt will be too steep for the air to flow over it and down, preventing the boundary layer from forming. When this happens, it's called a stall.

The second physics effect that causes an airplane to fly is also related to the angle of attack, and it's where Bernoulli's principle actually comes into play. The angle of attack causes air to flow more slowly and pile up under the wing. On top of the wing, only the air closest to the wing sticks to it. Air that doesn't get sucked into the boundary layer goes over the top of the wing and continues on its merry way. This means there is less air behind the wing than in front, creating a vacuum. The combination of high pressure and air pushing up on the bottom of the wing creates lift. Because the angle of attack is key to creating lift, pilots control the altitude of a plane both by changing the amount of force exerted by the engines and by adjusting the angle of attack. An angle of roughly 15° gives maximum lift; angles higher and lower than that create less lift (figure 7.1).

What does all this mean in terms of hard numbers? The lift force is dependent on the density of the air around the wing, the speed at which

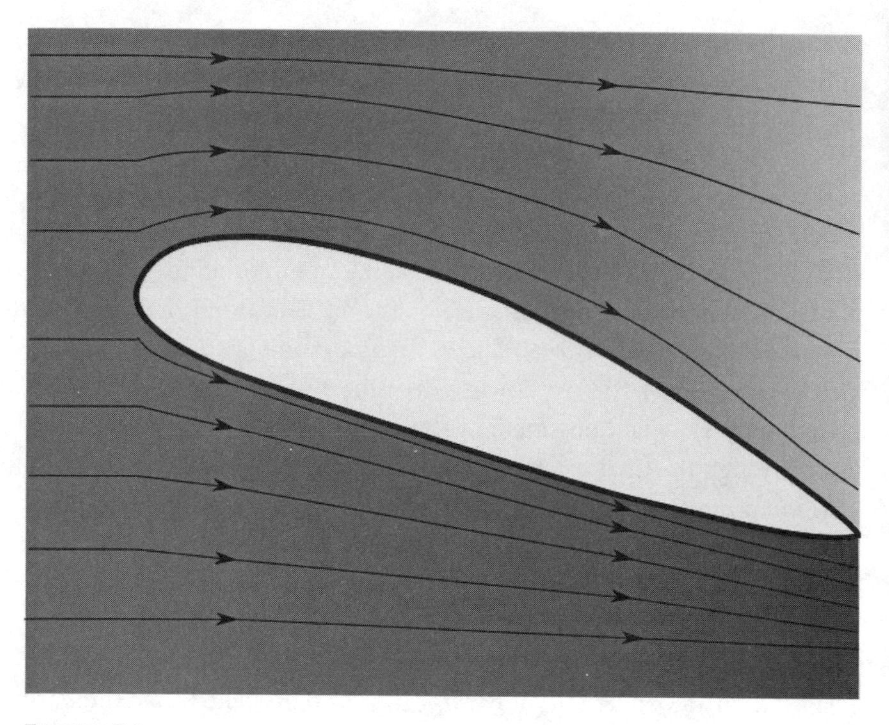

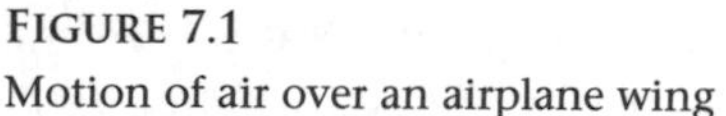

FIGURE 7.1
Motion of air over an airplane wing

the wing is moving, the area of the wing, and a parameter that takes into account the shape of the wing, the angle of attack, and how "sticky" the wing material is. This parameter is called the lift coefficient. In math terms, lift is calculated as

$$L = \frac{C_L \rho U^2 A}{2},$$

where ρ is the density of air, U is the velocity, A is the area of the wing, and C_L is the lift coefficient. We can apply this to the specific case of a 747. A 747 without fuel weighs about 565,000 pounds. This means the lift force must equal 565,000 pounds, or about 250,000 N, to get the plane off the ground. How quickly does it have to move to take off? We can use the formula to find out. An average lift coefficient is about 2. The density of air is 1.22 kg/m^3, and the area of a 747's wings is about 510 m^2.[2] Plugging that into the lift formula, we can see that the plane would have to be moving at about 63 m/s (140 mph). A 747 usually flies much faster than that, but as long as it's going at least 140 mph, it won't fall out of the sky.

Knowing how an inanimate Cessna gets off the ground may not seem like the key to understanding how a dragon might do it, but it is the first step. Birds are a bit like flapping 747s. Airplanes get air flowing over their wings by using propellers or jet engines to move the plane forward at high speed. Birds don't have jet packs strapped to their backs, however, so they have to move forward, or create thrust, under their own power. Birds are extremely light relative to their size, so they don't have to create that much lift to get their tails in the air, which means they don't have to create that much thrust, either. The lift equation shows that wings with more area will give more lift. Birds want to maximize the lift-to-weight ratio, and they do that with feathers. Feathers create a huge wing area without adding extra weight to be lifted.

BIRDS AND BATS AND ALL THAT

Bird wings are similar to plane wings once air starts flowing over them, but getting them going is hard. There are two motions to a flap: up and down. You've probably already guessed that most of the lift and thrust comes from the downstroke. When a bird's wing flaps downward, it creates forces in two directions: up and forward. It's easy to see how it can create upward force; it's pushing the air downward. So where does the thrust come from? Unlike a 747, a bird can change the angle of its wing relative to its body. On the downstroke, a bird will tilt the leading edge of its wing downward. When the wing is tilted forward, the push of the wing is going both down and back, not just down. This is similar to a how a swimmer moves in water. If you look at Michael Phelps' hands as they move through the water, they are tilted just like a bird wing. Think about how Phelps moves when he swims the butterfly stroke. He is going both forward and up out of the water, just like a bird . . . only wetter.

A lot of interesting physics happens on the upstroke, too. Because a bird can't get much lift or thrust from the upstroke, the goal becomes minimizing drag, or the force required to move through the air. Because a bird wing is hinged, it's possible to retract the wing a bit on the upstroke, reducing the wing area and creating less drag. The primary cause of drag on the upstroke is the feathers. On the downstroke, the feathers are turned so they are perpendicular to the direction of motion to create a wing with the greatest surface area, which in turn creates the most lift and thrust. On the upstroke, some types of birds turn their feathers sideways so that the air passes right through the wing. This neat little adaptation means much less energy is wasted pulling the wing back to a place where it can be useful again.

Takeoff requires the most energy from bird (or bat or pterosaur, which we'll talk about later). Not all that energy has to come from wing beats, however. Birds often run or jump in the air to add a little bit of lift and thrust to what their wings give them. Africa's largest flying bird, the kori bustard, weighs about 45 pounds, so it would need to create about 196 N of lift force to get off the ground. This means it would have to be moving forward at a speed of about 16 mph to get off the ground through aerodynamic lift alone. This is one reason they spend most of their time on the ground. When they do want to fly, however, they generate the speed needed for takeoff by running really fast into the wind. For a bird that huge with wings that aren't quite big enough, there aren't other options.

Once a bird is in the air, most use a combination of flapping and gliding. Gliding is a much more energy-efficient way to fly, and as long as they are going faster than the needed speed for lift, they can keep gliding. It's a bit like riding a bike; once the wheels are moving, if you aren't in much of a hurry, you only need to pedal every now and then to stay upright. Birds can change the shape of their wings and the angle of attack to control their height and forward speed. They can also use wind, air currents, and thermals to stay aloft. Hot air rises, and when it rises up from the ground it's known as a thermal. They are usually created over surfaces that easily absorb the sun's warmth, such as dark rocks or asphalt. Birds can ride these columns of warm air into the sky. For some, this is their main method of generating lift. Hold on to the concept of thermals—it's important when we talk about how dragons might be able to fly.

As a side note, I can't move on without addressing the question that has plagued nerds for ages: "What's the airspeed velocity of an unladen swallow?" Both African and European swallows cruise at about 11 m/s (24 mph), but they would only need to be moving at about 2 m/s (4.5

mph) to take off. I'll leave the question of the addition of a coconut as an exercise for the reader.

Bats and birds both flap and fly, but that's about all they have in common. One is a mammal and one is a bird, one has fur and one has feathers, and despite Alfred Hitchcock's best efforts, I am only terrified of bats. There is also, unsurprisingly, a huge difference in the way their wings are constructed. Dragons have bat wings instead of bird wings, but why? Many reasons, but probably mostly because they are terrifying.

Bird wings consist of a long arm, much like a human arm, but with only 2 fingers at the end. This makes up the leading edge of a bird wing; the rest is just feathers. Bird bones have the same stiffness throughout the bird and are porous, like a sponge. Bats also have an arm bone as the leading edge of the wing, but that's where the similarity ends. Bat bones are not hollow and they become more flexible toward the edge of the wing, and they obviously have a membrane wing instead of feathers. Bats can carry up to 20% of their mass in their wings as opposed to birds, which carry almost all of their weight in their body. One of the most important differences is that they have five finger bones that spread across the wing instead of two at the tip. With that many fingers, they have a lot more control over the shape of their wings. Bats can fly more precisely and are more efficient than birds are at conserving energy during long flights. They can perform some amazing maneuvers because they are able to so accurately change the amount of lift and thrust in their wings by moving their fingers. A bird would not have the ability to land upside down (figure 7.2). Also, like dragons, bats can walk on their wings and even swim with them. They are much more versatile creatures. (Fun fact: if you turn a picture of sleeping bats upside down, they look like they are having an awesome dance party.)

Because of the bat wing's bone structure and the flexibility of its membrane, the flapping motion of a bat is much more complicated than that of a bird. Bats change the shape and angle of their wings with more dexterity than birds are capable of—so much so that understanding what unique and efficient fliers they are is essential to developing the next generation of flying machines, such as drones. One of the foremost bat research labs is headed by biologist Sharon Swartz and engineer Kenneth

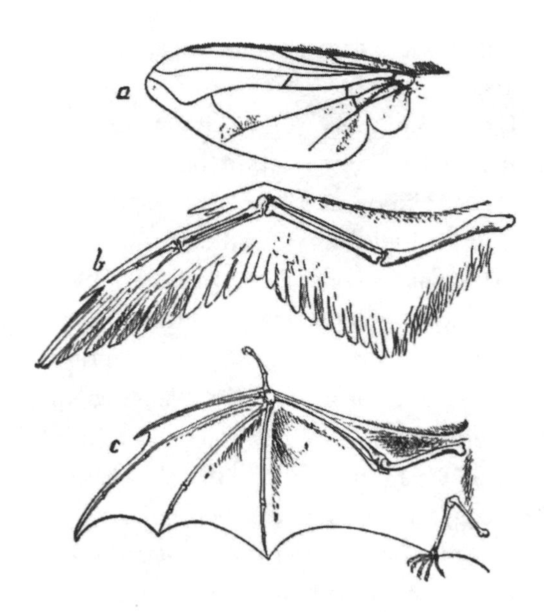

FIGURE 7.2
Comparison of insect, bird, and bat wings. Notice the bat wing looks like a human hand.

Breuer at Brown University in Providence, Rhode Island. Using technology very similar to that used in CGI animation, they've created a computer model of bat flight to use as a reference in their quest to build a robotic bat wing. If you have time, I would highly recommend checking out their research group's website and watching all the videos—slow-motion videos of bat flight are amazing.[3]

It is tempting to assume bats simply flap up and down and that they fly just like birds: an upward angle of attack on the downstroke to create thrust and lift and a retracted wing on the upstroke to minimize lift, with the only difference being bats' ability to better control the angles. In reality, the motion of a bat wing during flight is starkly different. If you had the chance to watch the slow-motion video, you might understand why Dracula can turn into a bat. The wing stroke is reminiscent of a vampire (a traditional vampire, not modern ones such as Spike and Angel) flapping its cape. On the downstroke there is a lot of forward motion, a bit like Michael Phelps trying to do butterfly in reverse or the windmill arm motion of someone trying to avoid falling backward. The leading edge of the wing is indeed pointed down as in bird flight, but as the wing moves down, it also moves forward significantly with the membrane dramatically billowing behind like a small cape. The wings move forward enough to almost cover the bat's face. This is exactly the opposite of bird wing motion, which looks like Michael Phelps swimming forward in butterfly stroke. According to Swartz,[4] this forward movement helps air flow across the wings to generate more lift.

The mechanics of the upstroke are still being debated. Because bats carry a significant portion of their weight in their wings, some researchers have theorized that if bats pushed backward enough on the upstroke, they could use the inertia of their wings to propel their bodies forward. Equal and opposite forces again—heavy wings go back, body has to go forward. After observing bats, Swartz concluded that there is no forward body movement during the upstroke.[5] A group from Sweden begs to differ. Anders Hedenström from Lund University used smoke and mirrors—well, smoke and lasers—to look at what happened to the air after a bat flew through it. In contrast to birds, which leave a relatively simple wake, Hedenström found that a very complex system of vortices were shed as bats flapped.[6]

What is a vortex? When an object is moving forward in a fluid such as air or water, its motion causes disturbances in the fluid. These disturbances, or vortices, then continue moving through the fluid as they are "shed" from back of the moving object, traveling in the opposite direction from the force and motion of the object (in this case, the bat wing). By examining the vortices shed during different phases of a bat's wing beat, Hedenström showed that vortices were shed during the downstroke and the upstroke. After analyzing the movement of the air, he concluded that the upstroke produces thrust and lift. How that thrust and lift is created differs depending on the speed of the bat. In general, bats create these forces not through inertia as previously thought but by inverting their wings. At low speed, the wing moves up and back and the leading edge is flipped almost 180° on the upstroke. It's like a little flick of the wing tip. Think about Dracula dramatically throwing back his cape before biting a neck. At faster speeds, the wingtip doesn't flip over but stays more vertical and creates much less thrust and lift. There's not a sudden cutoff speed where a bat goes from flicking to not flicking but rather a smooth transition from more to less flick. It's nothing like how a human transitions from walking to running.[7]

In addition to all the cool stuff already mentioned about bat wings, they can also help protect a bat during collisions. This is where inertia really does play a part. Inertia is the resistance to motion. Heavy things are harder to move than light things, and some configurations of mass are harder to move than others. An object's resistance to motion is called the *moment of inertia*. (This is where I apologize to all my Intro to Physics students for making them learn a completely different definition of the word "moment.") Because bat wings are so heavy, they really don't like to move much when they encounter an opposing force. If a bat is flying and gets hit in the head, its body will roll, but the heavy wings will stay put and help the bat right itself. Unlike birds, whose wings react easily to forces, it's much easier for a bat to keep flying when hit. The ability to remain airborne would be very important to a dragon in battle. A rock hurled from a catapult won't knock it out of the sky.

Given that dragons have batlike wings and wing mechanics, it's a natural question to ask how big bats can get. Is there a reason they are not the size of dragons? If you want to have some truly unique nightmares, I suggest watching some videos of the giant golden-crowned (Viserys would not be amused) flying fox, the world's largest bat. It has a wingspan of almost 5 feet and weighs roughly 4 pounds. Though it eats fruit and not blood, you couldn't pay me enough to get near one. Think about that for a sec—an animal weighing only 4 pounds needs just a 5-foot wingspan to fly. The dragons in season 7 are roughly the size of a jumbo jet. Is that even possible?

Dinosaurs

If bats don't get to be dragon-sized, and neither do birds, then has there ever been an animal that is both dragon-sized and able to fly? For those who went through a dino-loving phase (and who are we kidding—that was all of us), I'm sure your mind immediately jumped to the terrifying pterodactyl. I have bad news for those readers who, as children, dreamed of having a pet pterodactyl. Brace yourselves: pterodactyls did not exist. Flying dinosaurs *did* exist, but as a group they are categorized as pterosaurs, with no one dinosaur bearing the name "pterodactyl." I'm truly sorry. But they offer the most promising explanation of dragon flight. Weighing in at up to a half ton with a wingspan of over 30 feet, these creatures were massive—and possibly flighted. There is much debate in the paleontology community as to which of the of the pterosaurs were flighted and which were too massive to fly. If it's not something you can see, it's definitely something you can argue about.

Pterosaurs come in a huge range of sizes, from the tiny *Nemicolopterus crypticus*, with a wingspan of only 10 inches, to the giant *Quetzalcoatlus northropi*, with a wingspan even bigger than Drogon's at 36 feet. *Q. northropi* is estimated to have reached a weight of up to 250 kg (550 pounds), about the size of two baby elephants. For a long time, it was assumed that pterosaurs this big were not able to fly. Similar to the ostrich or emu, their wings could not support their body weight. It seems impossible that something the size of two elephants could get off the ground with only thin membranes for wings. But what if they were skilled at running and jumping off of cliffs, catching thermals, beating their wings occasionally, and gliding efficiently? Would they have been able to fly then? Surprisingly, the limiting factor in pterosaur flight isn't its wing membrane or whether it can create enough lift; it's whether their bones can withstand the force of flight. The wings of such a huge beast feel an enormous upward force to keep them aloft. Were the bones of pterosaurs up to the job? For years, scientists agreed the answer was no.

It's impossible to study live dinosaurs, so paleontologists have to compare them with nonextinct animals that share some similarities. In the case of the pterosaur, it's easiest to compare them to birds. Like birds, pterosaurs had light, hollow bones. Their wings were similar to birds' in that there is a leading arm bone ending in fingers that do not extend into the main part of the wing. Birds have two fingers to the pterosaur's four. The biggest difference seemed to be the difference between feathers and membranes. Because of the similarities between the two, researchers often used bird bones to test theories about dinosaurs. This worked well in the majority of cases, but as is often true in science, the cases in which things don't go to plan are often the most interesting. If you assume *Q. northropi* has bones that are the same strength as those of birds, it would be as grounded as an emu. The force needed to keep such a huge beast in the air would immediately cause his bones to snap if they were identical to bird bones, so paleontologists assumed for years that although they had the structure and adaptations to prefer a life in the air, pterosaurs could not get off the ground.

Enter Mark P. Witton from the University of Portsmouth, UK, and Michael B. Habib of Chatham University. They published a paper in 2010 supporting the idea that even the huge *Q. northropi* might have been able to fly.[8] It would have been easiest if they'd been able to take *Q. northropi* bones, load them with weights, and see how much it took them to break. Unsurprisingly, they were not allowed to do that. What they did do was take measurements of the bones and use those to determine the bone strength. Pterosaur bones are similar to human bones in that they are kind of like PVC pipe; both have an outer radius and an inner hollow section. To determine the strength of a pterosaur bone, Witton and Habib measured a cross section of bone to see how thick it was. They assumed that the breaking point of a solid pterosaur bone would be similar to the breaking point of a bird bone (175 MPa of force). Once they determined the strength of the bone if it were solid, they subtracted out the hollow core. They used CT scans of existing pterosaur bones to figure out how wide the hollow center was. They found that, due to their size and structure, the bones were up to three times stronger than bird bones. In the case of *Q. northropi*, the bones would have been about one-and-a-half times stronger than what was needed. So, it is indeed possible for massive creatures—maybe even dragons—to fly. In fact, the experts themselves didn't even rule it out, closing their paper with this sentence: "In all likelihood, there is no universal maximum for any major characteristic, including size, that can be applied to all flying vertebrates, or even most of them."[9]

FINALLY, DRAGONS

Now, what about dragons? Thanks for bearing with me through long explanations of flight in everything *but* dragons. It's important, however, to understand how both animals and 747s fly to understand how an animal the size of a 747 can fly. When thinking about fantasy animals such as dragons, it's usually a good idea to think to yourself, "If this animal is so cool, why didn't nature think of it first?" That's a good place to start when it comes to dragons. Why don't we have huge flighted animals? Can't we just scale up a bat—both in size and, well, actual scales—and say it works? The reason it's impossible to simply scale up flighted animals is because weight increases as volume and wing size increase, but the ability to create lift increases with area. Doubling the size of a bat would result in an eightfold increase in weight but only a fourfold increase in the area of the wings. This leads to a lot of interesting problems. Would it be possible for an animal to overcome these physical limitations and become a huge airborne flamethrower?

Let's start with what we know. According to GRRM, Drogon has a 20-foot wingspan when he enters the fighting pits. If we assume they have the same aspect ratio (ratio of wing length to width) as bats,[10] this would mean his wing area is about 15 m^2. In the TV show, he looked to be about the size of an elephant, so let's assume he has a weight of 13,000 pounds. This means he would need 57,900 N of lift to stay off the ground. Though the lift coefficient changes continually during a wing flap, we can estimate an average of about 2.[11] This might be a bit too conservative of an estimate because bats are extremely adept at changing their wing shape to create lift, but it's a good starting point for doing the math. Plugging these values into the lift equation, this means that Drogon would need to maintain a speed of about 125 mph to cruise without the added lift from flapping. With the additional lift, he could go even slower and be fine. Considering bald eagles can create enough thrust to glide at about 95 mph, Drogon should have no problem staying aloft with his wing beats. Dragons probably have huge lungs because they need lots of air to breathe fire, so it's possible they weigh considerably less than an elephant, but I don't have a good way to estimate the lung size, and therefore weight, of a dragon. I'll go into possibilities in the next chapter, but for this estimate it's best to just assume the weight of an elephant.

Seeing as Drogon can breathe fire, he has another built-in mechanism for generating lift—his wings don't have to do it all. When Drogon sets things on fire, he creates heat, and hot air rises. If Drogon can strategically set things on fire, then he doesn't have to eat so many sheep because he won't need to expend as much energy to create lift. Birds and at least one species of bat use thermals for a large part of their flight. That lift is generated from heat-absorbing objects such as rocks and asphalt and isn't going to produce nearly as much lift as a burning fleet of enemy ships. Dragons are too big to use thermals from parking lots, but the considerably stronger thermals from burning armies would add to their aerodynamics and flapping lift. Calculating the force created by hot air rising from fire is extremely complicated, so I won't go into it here, but I feel comfortable in assuming it is enough to help lift a dragon. With this fire-generated lift comes a lot of turbulence. Air is being heated unevenly by spreading fire and the unevenly heated air rises unevenly, too. As a dragon is flying over its field of destruction it would be knocked around by the uneven currents. It's a good thing they have bat wings and not bird wings. The ability to very specifically change the shape of their wings to accurately control the lift and thrust is key to winning a fight and not falling out of the sky.

By making dragon wings similar to those of a bat rather than a bird or a pterosaur, GRRM and company gave it the ability to not only fly but also walk and even swim efficiently and effectively for a such a large animal. Because of bats' unique wing structure, they are able to walk with their wings, similar to a human walking on their wrists. GRRM's dragons can walk on their bat-like wings, making them as terrifying on the ground as in the air. Their bat wings also give them the ability to swim should they fall out of the air and into the Narrow Sea. Creatures with bat wings are two-legged creatures that act like they have four legs.

So far, it seems like nature simply made a mistake in not creating real dragons. They could survive cold climates, move fast enough to get off the ground, use their own fire to create thermals to keep them in the air, and even swim. But, of course, there is a catch. In this case, that catch is bone strength. As with pterosaurs, the muscles can produce all the force needed for something to fly successfully, but if the bones aren't strong enough to handle it, then the animal isn't going anywhere. Nature reached that limit with giant pterosaurs, which had extremely strong and light bones. The largest pterosaur could handle roughly 225 N of force on each square meter of wing. I know that there is a different amount of force acting on the tips of the wings than on parts closer to the body, but because this is a rough estimate and a short book, I will assume the force is the same across the entire wingspan. Based on the work discussed earlier in the chapter, *Q. northropi*'s wings fail at about 400 N/m^2. Their bones are almost twice as strong as they need to be, but would similar bones be strong enough to hold up a dragon? Keep in mind that Drogon's wingspan is described as shorter than that of the *Q. northropi* and he is also about 10 times as heavy, so don't be too hopeful.

Assume, again, that a dragon weighs the same as an elephant (57,900 N). With a wing area of 15 m^2, that means there is 3,900 N of force acting on each square meter of Drogon's wings. If his wings were the same as a pterosaur's, they would snap before he got off the ground. It's going to be hard to find bones strong enough in nature because evolution tends to go only as far as it needs to, and no animal needs bones that strong. It's also difficult to compare this to animals such as rhinoceroses or elephants because no one is researching how their bones might bend if they were to fly, and I couldn't get my hands on any elephant bones, anyway. So, it's safe to say that bones strong enough to allow a dragon to fly don't exist in nature.

When I initially set out to show how dragons fly, I assumed they wouldn't have been able to because of aerodynamic forces. They wouldn't be able to create enough thrust to create the lift needed to get them off the ground. I was as surprised as anyone to learn that, for the most part, they would be able to conquer Harrenhal by air. I would never have guessed that the limiting factor would be their bone strength. Dr. Jim Kakalios, author of *Physics of Superheroes*, often says he will grant each superhero one exception—one thing that isn't physically possible but makes the story work. I like this idea, so I think that in the case of dragons the one exception I'll give them is super strong bones. Let's pretend they are made with aluminum, or adamantium, or the mythical substance of your choice and get back to watching them rain destruction down on Westeros.

8

How to Kill a White Walker

The Physics of Dragonglass

When he opened his eyes the Other's armor was running down its legs in rivulets as pale blue blood hissed and steamed around the black dragonglass dagger in its throat. It reached down with two bone-white hands to pull out the knife, but where its fingers touched the obsidian they smoked.

—Samwell Tarly fighting a White Walker, *A Storm of Swords*

Valyrian steel can make short work of a White Walker, but there aren't many Valyrian steel blades left in Westeros. This presents a problem for the entirety of Westeros, especially with winter's approach and the problems from the north. Samwell Tarly was lucky enough to accidentally figure out that dragonglass can also kill White Walkers, too. It seems that the only things that can kill the ice beings is something made by dragons. Considering the book series is titled *A Song of Ice and Fire*, this isn't exactly shocking. Luckily, Daenerys believes Jon and lets him mine her glass, so to speak. Physics says steel won't work to kill a White Walker, as Sam found out. But, dragonglass—or obsidian, as it is known in both our world and Westeros—will do the job well. Why? What makes dragonglass and its real-life counterpart so special?

As GRRM's name for it would imply, obsidian is a type of glass. In general, glasses are shiny, often transparent, smooth, and tend to feel cold. The most common type of glass—the kind we interact with on a daily basis—is made from silicon dioxide (SiO_2), which has the same chemical formula as sand. Sand isn't exactly super shiny, and it is definitely not transparent. You might assume that it's something about the way the atoms are ordered in window glass that gives it its unique properties. If so, can anything be turned into a glass? How does the arrangement of atoms make an object either opaque or transparent? What does the crystal structure of glass look like? What makes some glasses dark and others clear? How is it supposed to help us defeat the army of the dead? So many questions about glass! You'd think asking a scientist would help answer some of them, but seeing as scientists can't even agree on whether glasses are liquids or solids, that might be tricky. Hopefully, I can at least shed some light on (or through) the science of glass.

So far, I've talked about other solids: rock, ice, steel, and Damascus steel, to be exact. All of them have a defined crystal structure; that is, the atoms sit in a particular order throughout the crystal. A solid's properties are macroscopic representations of what the atoms are doing at the microscopic level. The BCC and FCC lattices and metallic bonds are what gives metals their special properties—ductility, conductivity, and malleability, to name just a few. To understand glass and, more specifically, how it might operate at cold temperatures, it's important to start by understanding what glass looks like at an atomic level. Glass is what's called an *amorphous solid*, which means it doesn't really have a structure. It's basically as if the molecules played freeze tag—one moment they are freely running around in a liquid state, and then next thing they know they are all stuck in one place. It's a solid, but there's no long-range order. The atoms and molecules have stopped moving, but it happened so suddenly that they haven't had time to organize themselves. This type of arrangement is very similar to a liquid, which is why sometimes people see glass as a solid and some see it as a highly viscous, slow-moving solid. There's even a very famous experiment, which I'll discuss later in the chapter, that shows it can be a liquid. Hopefully by the end of this chapter you will have a greater appreciation for the stuff you've probably run into by accident while carrying coffee.

Solid, Liquid, or Both? What Is Glass?

Whether or not it's a good analogy, I always think of glass as a liquid that's playing freeze tag. It inevitably makes me think of third-graders at recess. If they are running around outside and are yelled at to stop, they'll stop exactly where they are in no particular order, just motionless disorder. If they are told it's time to head inside and sit down, they'll take time to line up and file into their classroom. Most liquids do the same thing, except instead of being told to stop, they are cooled down. If the cooling is done quickly, they don't have the chance to slowly form into orderly crystals like they want to. Remember from the chapters about steel that quenching creates a solid structure that is markedly different, and better, than the ones created with slow cooling. A solid is defined as having closely packed molecules arranged in a regular array—a rigid structure that does not deform or flow to fill a container or a specific volume and cannot be compressed easily the way a gas can. The solids I've talked about until now are called *crystalline solids*, meaning they have a crystal structure. An amorphous solid, by contrast, is one with no long-range order. It still has the properties of a solid (for the most part) but no one part has the same order as any other part. The word *amorphous* comes from a Greek word meaning "without shape." Although the final glass has a shape on a macroscopic level, microscopically, it does not (figure 8.1). Another way to understand amorphous solids at the microscopic level is this: Imagine you are a tiny observer sitting on a crystal in a crystalline solid. As you look around, you see a regular array of atoms. You say, "OK, there's sodium to my right, and down there I see chlorine." This gives you a sense of directionality and orientation within the solid. Now, put yourself inside an amorphous solid. The first thing you notice is that everything looks the same in every direction. There is no structure to orient yourself around, to the extent that if you closed your eyes and took five steps to the left and opened your eyes again, your surroundings would give no indication that your location had changed. This gets at the more mathematical definition of amorphous, meaning that there is no geometrical reference frame for defining motion within the solid.

The study of glass has a long history. Just like steel, it was first made about 5,000 years ago. The first scientific papers on the properties of glasses were published about 200 years ago. Considering writing had barely been invented when glass was first studied, it's not surprising that it took a while for articles to be published. John Mauro and Edgar Zanotto published a fabulous history of the study of glass over the past 200 years.[1] They found that the rate of publication of papers on glass has increased exponentially since 1945. More recently, the number of patents

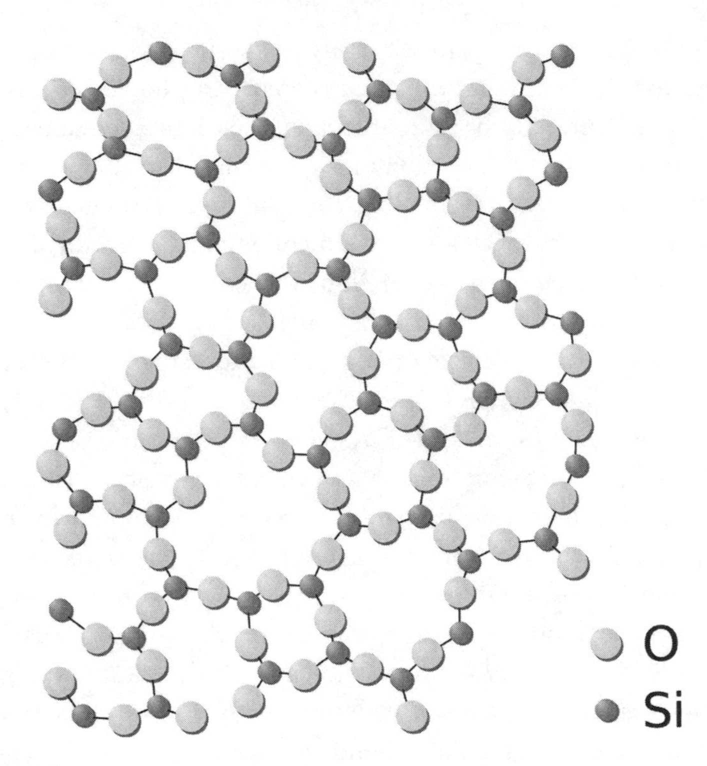

FIGURE 8.1
Example of the structure of a glass. This is silicon dioxide, the primary component of window glass.

has surpassed the number of papers published. As you will see in chapter 10 when I talk about Greek fire, there's a difference between science and technology. With more patents being issued, glass has moved from an area of fundamental research to a focus of really impressive technological advances. The glass on your smartphone is just one amazing example. Automotive glass is another precisely engineered technology.

Many different types of materials can be made into glasses and the method is generally the same: heat something until it melts and cool it quickly enough that it doesn't have time to crystallize. Just as different materials have different melting points, they also have different rates of cooling required to freeze them into a glass. For something like quartz, which is composed of silicon dioxide (also the main ingredient in window glass), once it's been heated to a certain point, it can be cooled in air and become a glass. It's almost impossible to turn it back into a crystal. A substance such as steel can be pushed into a glass phase by being cooled at a rate of about 1 million degrees per second. If people really wanted to make a fabulous sword, metallic glass would have been the way to do it. Compare that to the cooling rate of quenching a sword, which is about 1,000 degrees per second, maybe less. This is why steel, even if heated until molten, doesn't become a glass when quenched. The temperature at which something changes from a molten state to a glass is called, unsurprisingly, the glass transition temperature, or T_g. One question that arises is how to determine when something becomes a glass. T_g is different from the melting point, which is where a material would go from a liquid to a crystalline solid. With a melting point, there is a quick jump from ordered solid to disordered liquid. Because a material sort of slides into glass form, there's no quick and easy way to determine the point at which it becomes a glass. There is a huge question of exactly when something goes from liquid to glass on a microscopic level. In a crystalline solid, it's pretty easy to see when the atoms and molecules are all sitting in an ordered way. In a glass, there's just a point where the molecules stop moving enough to be called a glass. There are different ways to determine when this point is reached. The most common involves measuring how quickly the molecules are moving or how stiff the material is. Once the molecules are moving slowly enough and the material is stiff enough, it's considered a glass. Glasses are kind of like obscenity—hard to define, but you know it when you see it. It's worth noting that the only way to produce a glass is to heat it and quench it into place. Nothing has a ground state as a glass. The atoms don't usually choose to line up in an amorphous disordered state.[2]

I know I defined glass as an amorphous solid and that I have been using the two terms interchangeably up to this point. Many people do that, but some scientists are picky about it. Not surprisingly, however, they aren't always picky in the same way. A paper by P. K. Gupta does a great job of summing up the various definitions of glasses versus amorphous solids.[3] A glass is defined as something where the order over a very short range is indistinguishable from that of the same material in a liquid state. An amorphous solid is one that doesn't have long-range order or crystal structure, but it is possible to tell the difference between a liquid and a solid in the short range. This is probably more useful to know for advanced bar trivia than for this discussion, but it's still interesting.

One of the hallmarks of glass, as anyone who's managed to drop a glass of wine at a cocktail party knows, is that glass is brittle and shatters easily. Going back to the terminology of chapters 5 and 6, glass is not very tough and not at all ductile. What makes metal so ductile is the crystal structure that allows the atoms to roll over each other when pushed. With too much carbon in the way, this rolling is impeded, and the material becomes brittle and prone to fracture when forged. When it breaks, it does so along dislocation planes, and oftentimes the grain boundaries act like little roadblocks stopping the cracks from progressing. The metal can slip and roll and move without structurally failing. In glasses, however, there's no order at all. The atoms can't roll over each other when pushed, and there are no easily defined fracture planes. When pushed, the molecules have nowhere to go and are under a huge amount of stress, so they crack. Because there is no long-range order, there are no roadblocks to stop the crack from continuing. The material simply shatters. And, in contrast to crystalline solids, there typically is not an ordered pattern to the cracking. You'll notice that when a crystal shatters, the small pieces look similar to the original pieces; after all, little rocks are just tiny versions of big rocks. But when glass shatters, you see a whole variety of shapes and sizes: some tiny, deathly sharp needles; some big chunks that are easy and safe to pick up.

Glass can be strengthened through a process called *tempering*, which I mentioned when talking about regular steel. Tempered glass is about four times stronger than normal glass and has the added benefit of shattering into small, regular pieces with low stabbing possibility. Tempering glass is very similar to tempering steel. It is heated to just below the melting temperature and then quenched with air. The outside cools much more quickly than the inside. Unlike with steel, where the process results in a different form of steel on the outside from the one on the inside, the whole thing remains glass because glass doesn't really have any structural or crystal phases. The inside cools much more slowly, however, and as it cools it pulls back on the outside, causing stress. This makes for stronger glass, but when that tension is released after too much pressure is applied, the glass basically explodes. Pyrex is a great example of tempered glass.

It's the tempering process that allows you to bake delicious brownies in a Pyrex dish and stay worry-free if you drop the hot dish on the ground or run cold water on it before it cools.[4]

One of the most beautiful examples of glass is the stained-glass window. The development of both glass and steel happened around the same time and both followed a similar path. Initially, artisans would use whatever material was available to create glass and, as with steel, it came out however it came out: sometimes it worked, sometimes it didn't. Glass could be made in a huge array of colors depending on the materials used to make it, and eventually it became clear as to which materials would produce certain colors. The colors come from trace minerals or elements mixed in with the raw materials of the glass. Iron oxide, for example, will tint glass green, whereas cobalt will lend its characteristic blue hue. One of my favorite glass additives is uranium. It's not the safest, but glass with uranium mixed in glows spectacularly under a black light. (You can get uranium marbles on Amazon and see for yourself.) Many of these colors, however, will fade over time. For a 1,000-year-old church window, this can pose a problem. Ancient glassmakers stumbled on a solution, however: nanoparticles. Turns out artisans were pretty darn good at using them without knowing it. Adding gold and silver to glass created beautiful colors that didn't fade. One of the classic examples is a cup from Rome dating back to the 13th century AD. The cup appears green at first look but red when light shines through it. Scientists imaged the glass and found that it contained gold and silver nanoparticles. The mechanism by which these particles were created is still unclear, but medieval artisans knew how to use them to their advantage. When light hits the nanoparticles in the cup, it excites the electrons on the surface of each tiny little ball. In a sheet of gold, the light is reflected normally and the sheet looks gold, but with nanoparticles, the electrons are so close together that they have very little room to move. They preferentially reflect red light, making the nanoparticles, and thus the goblet, appear red.[5]

There are several examples of glasses being created in nature as well as by accident. One type of natural glass is *fulgurite*. When lightning strikes sand, it is hot enough to melt some of the silicon dioxide, which then solidifies back into a glass. Glass is created along the path the lightning takes through the sand. This creates beautiful pieces of glass that look like glass branches. Humans accidentally on purpose did something similar at the Trinity nuclear test site in New Mexico. During the first nuclear bomb test, sand was drawn up into the fireball and melted. It was so hot that the molten sand rained down, cooling on its way to the ground and creating *trinitite*. Many scientists from that time still keep pieces of trinitite on their desks or have passed it down to their students. Trinitite is mildly radioactive, but not enough to harm anyone. Still, it is now illegal to remove it from the test site. Trinitite is generally light green in color and has a complex structure due to its unusual creation. Volcanoes can also create glass—obsidian, to be exact—but I'll save that discussion for later.

The Sad Case of the Pitch of John Mainstone

In 1927, Professor Thomas Parnell of the University of Queensland in Brisbane, Australia, walked into his physics class with pitch, a funnel, and the goal of showing his class that sometimes solids weren't solid but liquids that moved very, very slowly. He took a block of pitch, heated it, and put it into the funnel. Pitch isn't technically a glass but is a great example of something appearing to be a solid but acting like a liquid. That, and this is one of my favorite stories in science. Then he waited. And waited. After allowing the pitch to settle for three years, and probably watching some of the students in that original class graduate, he put a beaker under it and cut the tip of the funnel. Then he waited some more. Those original students graduated med school and law school and had babies. During all this time, the pitch was ever so slowly creeping out of the bottom of the funnel and forming a drop. It looked just like a water drop, only it wasn't moving on any scale we can see. As the drop of pitch got closer and closer to the bottom of the beaker, the experiment was monitored more and more closely. Parnell waited (and waited and waited) to for the moment when the pitch drop fell. As the drop slowly moved down, it eventually hit the bottom of the beaker but was still attached to the mass of pitch. "Falling" was when the drop officially broke that last connection with the pitch in the funnel. Though it took years for the drop to hit the beaker, the actual break would take only a fraction of a second. Eight years after the funnel was first cut, in December of 1938, the drop, which had been sitting on the bottom of the beaker, finally cracked off of the rest of the pitch. After eight years, the actual drop took only a fraction of a second. No one saw it. A second drop fell within Parnell's lifetime in 1947, and he did not see that one either. The third drop fell in 1954, after Parnell's death and before anyone else seemed to care much. In 1961, John Mainstone joined the faculty and became the custodian of the experiment (figure 8.2).

FIGURE 8.2
Pitch drop experiment setup. You can see the large drop of pitch inching ever closer to the bottom.

By this point, 13 years after Parnell's death, the experiment was shoved in a cabinet. Mainstone wanted to watch and hauled it out from obscurity. Another drop had fallen sometime in the intervening years, bringing the count to three drops in 31 years. Mainstone was determined to see the drop fall. He didn't have long to wait. A fourth drop fell in 1962. He missed it. Another in 1970. Though Mainstone watched religiously, he missed it once again. He was so close in 1979—he looked at it on a Friday, came into the office to monitor it that Saturday, and decided that it would hold off on breaking until Monday. It did not, and Mainstone missed it. *Again*. He was determined not to miss it in 1988. He kept vigil over the drop but, as one does, he needed a cup of tea and walked away for a minute. After nine years, the drop chose the moment Mainstone wanted tea to break. No one had ever recorded the fracture of a material such as pitch, and Mainstone desperately wanted to see what would happen at the exact moment a drop broke away from the bulk material. As the next drop was getting close in 2000, he decided to use modern technology to capture the elusive moment of fracture. He set up a camera to record the event because, darn it, he didn't want to miss it because of another tea break. He left the camera recording and went on a trip outside of the country. His grad students emailed him saying the drop was about to go, but he was confident it would all be recorded. I remember being a grad student and the crippling fear I felt when emailing my advisor to say something wasn't working. I would not have wanted to be Mainstone's grad student on November 28, 2000, because the camera had malfunctioned. After 12 years, the camera had failed to catch the latest drop. You may notice that the time between drops increased between drops four and five. This is because an air conditioning unit was installed, slowing the flow of the pitch by roughly four years. Pretty impressive air conditioner.

It began to look like the drop would fall again in 2013. There was quite a bit of press coverage of the experiment and of Mainstone's story. A live feed was set up with three cameras. Many people had a "pitch watch" window open in their web browsers. It was so close and with so many eyes on it, someone was bound to see it. The tech seemed to be working perfectly and there was triple redundancy. In an interview for Radiolab, Mainstone seemed thrilled by the prospect finally seeing the drop break. On April 24, 2014, the drop finally fell, with the world—or three cameras, at least—watching. It was caught on film for the first time eight months after John Mainstone died.

Obsidian

In the books, the terms "obsidian" and "dragonglass" are used interchangeably. Dragonglass is not something I'd try to buy on eBay, but obsidian is not terribly rare. It is unclear if dragonglass was obsidian made with dragon fire or if it was produced in a volcanic explosion. Some say it came from dragons; some say it came from the earth. Either way, it's pretty darn cool. GRRM has assured us that there is an element of magic to dragonglass, but as we saw with Damascus and Valyrian steel, there may be a whole lot of physics in it, too.

Obsidian is formed when molten lava is cooled quickly, just as you would expect for a glass. The same magma that produces granite will produce obsidian. Both granite and obsidian are *igneous* rocks, meaning they are formed from cooling magma. They have roughly the same chemical composition but very different structures. The key difference between the two is the cooling time. Granite, an *intrusive* igneous rock, is formed under the Earth's crust and cools very slowly, giving crystals plenty of time to form. Obsidian, however, is an *extrusive* igneous rock. Extrusive just means that it got spit out before forming. Obsidian is formed from the magma that leaves the Earth's crust and granite is formed from the magma that stays inside. Because the transition from molten rock temperatures to surface temperatures causes some pretty quick cooling, obsidian forms. It usually only forms on the edge of lava flows, though, because the areas around the center have some insulating effects. It is most often black due to inclusions from the various elements in the magma and the formation of small crystals during the cooling process. Sometimes neat bubbles will form and, in the case of snowflake obsidian, sometimes the crystal pattern is visible to the naked eye.

Obsidian has been used as a weapon for thousands of years. It was highly prized in the Stone Age for its glass properties. It is one of the few naturally occurring glasses and it is by far the most abundant. Early civilizations found that when obsidian broke, it did so in a way that left a very sharp edge. For those who were less into stabbing things, it could be used to make mirrors and jewelry. Many glasses have the same useful properties, but by the time people learned to make their own glass, their civilizations had already upgraded to bronze as their material of choice for weaponry, and they were well on their way to producing steel. In 1975, an anthropologist, a materials scientist, and a geophysicist teamed up to write the definitive paper on the properties of obsidian.[6] They took 28 samples of obsidian from all over the world and X-rayed them, stretched them, heated them, dipped them in acid, ran electrical currents through them, and scratched them to try and understand every last one of obsidian's properties. They then performed the same tests on granite, Pyrex, and normal glass to measure how the obsidian held up by comparison. As was expected, the team found that obsidian had a crystal composition similar to that of granite, and it also displayed a silicon dioxide content similar to that of Pyrex. Pyrex, however, did not have the array of trace elements present in obsidian.

Reading this paper, it became clear that this was turning into a race between obsidian and Pyrex as to which could stand up to the elements better. It's worth following the researchers' method because I'm guessing most readers have a working understanding of how useful and strong Pyrex is. Hopefully, the comparison will give you a better understanding of how cool obsidian is. The melting point of obsidian is much higher than Pyrex's, and it can withstand far more heat than Pyrex before softening. (Obsidian brownie pans are starting to sound pretty good.) When a glass is heated, like most other things, it expands. Obsidian expands about twice as much as Pyrex does for every degree (°C) the temperature is increased. Next, they threw Pyrex, obsidian, and window glass in hydrofluoric acid and measured how their weight changed over time to determine how much had been dissolved. Obsidian dissolved about 50% faster than Pyrex, but seeing as Pyrex barely dissolved at all, this is a really small increase. Window glass, by contrast, dissolved three times as fast as Pyrex. I've been talking about glasses as if they are an either/or situation—either the solid crystallizes, or it freezes into a glass. However, that's not really true. It is possible for a glass to have some crystals within it. Obsidian is of that type. As I'll discuss further in chapter 9, granite, and therefore obsidian, is partially composed of feldspar. In the obsidian samples they imaged, the group observed feldspar crystals scattered throughout the glassy matrix. After looking at the hardness, which varied widely between samples, it was theorized that the crystals strengthened the obsidian. Obsidian is much like the other White Walker killer in that way. Again, microstructures are giving rise to macrostructures that kill. I know that was a bit of a laundry list, but it got some of the fundamentals out of the way for those interested.

Now let's get to the good stuff: Why does obsidian work so well as a knife, and what happens to it in the cold? Obsidian is very hard and, like anything else, hard things are also extremely brittle. Obsidian will break very easily. It's not going to make a good sword. The way it breaks, however is quite useful. Stone weapons are time-intensive to make and easily broken. As a glass, obsidian can actually be shaped and sharpened just by breaking it. Glasses such as obsidian don't fracture like crystalline structures. Instead, they undergo *conchoidal fracture,* meaning they fracture like a shell. Crystalline solids such as ice and steel like to break along specific fracture planes. Because there are a lot of fracture planes in any given solid and they might not all be going in the same direction, the breaks aren't always clean. With an amorphous solid such as obsidian, there's no preferred breakage direction. When glass is hit, the impact radiates out like ripples on water. If you are like me, you probably have at least one crack in your windshield from a rock on the road. Look at it carefully and you'll see that it has a fracture pattern that looks like the ripples on the water after you throw a rock in. The molecules are sitting pretty much the same way they would in a liquid like water, so that's not surprising. When a rock is thrown in the water, it creates a force that radiates outward, and the water reacts by moving in ripples. Similarly, when a glass encounters a force, such as a rock flying toward it from a gravel truck, it wants to move like ripples on the water, but it can't because it's a solid. Instead, it breaks the way it would have rippled. These fractures leave very sharp edges. Because there's no crystal structure that needs to hang together, in some cases obsidian can cleave down to a thickness of just a few atoms. It's this fracture pattern that makes a broken wine glass so dangerous. Obsidian obviously had many uses back when the only other option was granite, but its unique fracture properties are still being used today. A small number of surgeons are now using scalpels made with obsidian blades. Obsidian can be much sharper than the normal steel scalpels because of the conchoidal fracture. Because the blades are so much sharper than steel, surgeons feel they lead to faster recovery. There are a few huge downsides that mean these blades most likely won't be widely used. First, the blades are so sharp that you wouldn't feel a cut from the blade unless it was disturbed by something else, such as water. If a surgeon accidentally cuts their own hand during a procedure, it's best to know as soon as it happens. Second, the blade cuts through skin so easily that surgeons would need to retrain themselves to get a feel for how these new blades work so they don't cut deeper than they'd intended. Retraining muscle memory is not an easy thing to do. Finally, obsidian is very hard, and with hardness comes brittleness. Obsidian blades are great for cutting skin straight on, but if the surgeon were to knock the blade against something else from the side, it would break. This isn't a big deal on the scale of something like an arrowhead, but surgeons need to know when their tools have been compromised.

Of obsidian's properties, probably the most important one to Jon, Sam, and everyone else about to come face-to-face with the army of the dead is how it functions at low temperatures. Well, in terms of strength, it's not exactly doing great at room temperature. It's hard and it fractures pretty easily. I was not able to find studies on the strength of obsidian in extreme cold, but I was able to find a paper on how crown glass, made with SiO_2 and B_2O_3 (boron trioxide), behaves at low temperatures.[7] In 1957, three years after Parnell's third pitch drop fell, Kropschot and Mikesell ran some tests on how much energy was required to break the glass, first at room temperature and again after it had been exposed to liquid nitrogen (−196°C). They showed that this type of glass gets stronger the colder it gets. The duo doesn't identify a mechanism for why this happened; they merely reported that it did, in fact, happen.

Sam versus a White Walker, Take Two

At the end of season 7, we watched an army of wights and White Walkers make their way into Westeros. As I said in my chapter on zombies (chapter 4), the best chance a random peasant from the North has is to get the hell off the road and hide out. That said, Jon's people are coming back from Dragonstone armed to the teeth with obsidian. There is no definitive answer as to whether dragonglass came from the mountains of Valyria or whether it was forged by dragons, but I'd like to assume it was dragons. In the next chapter, I'll talk about what happens when a dragon goes up against granite. (Spoiler: the granite doesn't win.) Dragon fire is hot enough to melt sand and stone (I promise I'll tell you why if you keep reading), and there is a huge cache of dragonglass under Dragonstone, right where a bunch of dragons used to live. Given that narrative, I'm going to assume that dragonglass is made by dragons melting granite, which has the same chemical composition as obsidian, and then letting it cool quickly. The only other weapon that was able to kill a White Walker was one forged with dragon fire. To me, it makes sense that dragonglass creation involves dragons.

Armed with dragonglass, how would the army of Westeros hold up against the army of the dead? That's a hard question to answer. Dragonglass has a lot going for it. Unlike steel, it strengthens when it gets cold. It's still quite brittle and will fracture easily, but when it fractures, it does so in a way that keeps it sharp. Unlike steel, if it shattered while inside a White Walker it would still be quite deadly. I had someone ask me why normal swords couldn't just be covered in dragonglass, thus making the ultimate weapon. Unfortunately, that wouldn't work: the obsidian would probably fracture off quickly; it's next to impossible to bind glass to steel; and it would shatter if a White Walker were struck with the actual blade, leaving the hero defenseless. I think the biggest challenge would be making sure there's enough dragonglass weapons for everyone to quickly grab another in case theirs shatters. Surgeons have had a hard time adapting to the new obsidian blades because they were balanced differently and so much sharper than their normal scalpel, and I could see trained swordsmen having the same issue with dragonglass. Obsidian is lighter, so an obsidian sword would be balanced differently than a normal steel sword. It makes a much better knife or spear than a sword, anyway, so some weapons training would certainly be required. There could potentially be more collateral damage as well if people were to slice themselves with their own dragonglass by accident and not realize it until much later. If a surgeon can cut themselves without feeling it during surgery, what might fighters do to themselves in the adrenaline-fueled heat of battle? At the end of the day (or the beginning of the Long Night, depending on your point of view), the physics of dragonglass clearly shows that it behaves much better when battling a White Walker than normal steel. If the people of Westeros lose this war, physics had nothing to do with it. As for training with new weapons, the aim is always the same: stick them with the pointy end.

9

Harrenhal

Can Fire Melt Stone? Take Down a Wall?

And the greatest of them, Balerion, the Black Dread, could have swallowed an aurochs whole, or even one of the hairy mammoths said to roam the cold wastes beyond the Port of Ibben.

—Tyrion Lannister, *A Game of Thrones*

Balerion the Black Dread, along with his kin Vhagar and Meraxes, were responsible for conquering Westeros. Balerion was said to be so large a horse could have ridden down his throat. He could fly (see chapter 7), but most importantly he could breathe fire. Harrenhal was built to be an impenetrable fortress. Unfortunately, no one saw an aerial attack coming. The dragons swooped in, and old King Harren the Black didn't stand a chance. Harrenhal is described as having melted—not burned down, not crumbled, but *melted*. Now, stone seems pretty solid. We hear of 1,000-year-old churches and stone buildings from the times of gladiators; stone is not supposed to melt. But maybe it could. And if it could, was Balerion capable of that? There are a few questions to be answered here. First, could fire ever burn hot enough to melt stone? If so, would Balerion's white-hot flames be enough? We all know Dany wants to ride her three (now two, *sniff*) dragons to King's Landing and stake her claim to the Iron Throne, which was forged with Balerion's fire. But are Drogon and Rhaegal up to the task? Balerion is described as being huge. Does size affect a dragon's ability to produce fire? Does dragon flame develop in the same way as dragon size?

The most important question, however, is whether or not a dragon can produce fire in the first place. There are a whole lot of explanations on the internet fan forums about how a dragon might produce fire. There are also lots of arguments debunking those theories. What's missing, however, is some math based on real-world numbers. I'm hoping here to confirm or refute as many of the dragon fire explanations as I can. I will admit from the outset that I have probably missed something. I am sure there is an obscure scientific explanation for how it could happen. However, nature and evolution are really, really good at their jobs. They gave animals giant brains, gills, prehensile tails, and cute faces that make us want to take care of them. Breathing fire is right up there with the opposable thumbs on the usefulness scale. Sure, nature also gave us the appendix, but if there were a way to have given us fire, I'm guessing nature would have figured it out.

What Is Fire?

Before understanding dragon fire, how it might be made, and how hot it can get, it's useful to start by understanding out what fire *is*. This is a rather complicated question. You may remember that in 2012 the actor Alan Alda challenged scientists to come up with a good answer for it.[1] When he was a child his teacher wasn't able to explain flame, so as an adult he wanted to fix that. Fire and flame are quite hard to understand and even harder to explain, as many of the contest's entrants can tell you. The first question many ask on the road to understanding is whether flame is a solid, a liquid, or a gas. The ancient Greeks were adamant that it was an element all its own, and they weren't far off. Fire isn't any of those things. It's energy being released as heat and light during a series of chemical reactions. What you are seeing is the energy that's released when bonds break and recombine into other things. The color of the flame depend on what atoms and molecules are involved in the reaction and the amount of energy released.

Even though we say the wood (or gasoline, or paper) is burning, it isn't. The chemical reactions that we see as burning can't take place when something is in solid or liquid form. The first step in burning something like wood is getting it hot enough to produce vapor that can then burn. The *flashpoint* of a material is the lowest temperature at which enough vapor is produced to burn. In spite of its colloquial use, the flashpoint is not the point at which something ignites—it is the point at which it has the potential to ignite. The temperature at which it actually ignites is called, fittingly, the autoignition temperature. Take gasoline, for example. Its flashpoint is very low—very, very low, in fact; around −45°F (−42.8°C). This is why it's still possible to light gasoline in the winter and why it's so dangerous to store gasoline. The ignition point, however, is around 536°F (280°C). The terms *combustible* and *flammable* are often used interchangeably, but the first means capable of burning and the second means it can burn when exposed to a flame. Whatever dragons do, it's got to be flammable, not just combustible.[2]

When a fuel reaches the flashpoint and is then either brought to its autoignition temperature or comes in contact with a hot enough spark, a combustion reaction starts. A combustion reaction is an *exothermic* reaction, or a reaction that releases heat. Because lots of heat is being released, the temperature stays above the ignition point and the reaction continues. This is why fire is so dangerous—it's a self-sustaining reaction. The specific reactions are different depending on the fuel being burned, but in general, the burning material is being oxidized, meaning it's being combined with oxygen. Air contains roughly 20% oxygen in the form of two oxygen atoms bound together (O_2). During a combustion reaction, the bonds between the oxygen molecules break and the oxygen combines with atoms in the fuel, usually carbon and hydrogen. When wood is exposed to flame, for example, the oxygen molecules break apart and recombine with the carbon and hydrogen atoms present in the wood to produce water and carbon dioxide. Energy can be neither created nor destroyed, so in a chemical reaction, the system has to contain the same amount of energy when it finishes as it did when it started. There is a specific amount of chemical potential energy in the original molecules that make up wood, but there's much less in carbon dioxide and water. So, when the combustion reaction occurs, that excess energy has to go somewhere. The fire you see is that extra energy being released in the form of heat and light. The color of the flame indicates how hot it is as well as which chemical compounds are releasing that energy. I'll talk more about that in a bit.

This type of reaction also occurs in processes like rusting, another oxidation reaction, but in combustion it happens very quickly. To form rust, oxygen from the air breaks its weak double bond and recombines with iron. When this happens, a little bit of heat is released, but it happens so slowly that it usually goes unnoticed because there's not enough heat to reach the flashpoint. If you soak untreated steel wool in lemon juice to get rid of its protective coating and then throw it in a ziplock bag, you can see it rust very quickly and feel the temperature of the ziplock increase. The reaction still isn't happening fast enough to be called fire. If you touch a 9-volt battery to very fine, untreated steel wool (#00, available at any hardware store), however, the reaction happens very quickly and causes a fire. This is a great way to start a fire while camping and a terrible situation if you store steel wool in the same drawer as batteries. Different types of fuel, such as gasoline, allow combustion to happen much, much faster. The faster it can happen, the more explosive the fuel is. Controlling the combustion reaction has been integral to modern life. Gas lamps and candles burn slower than jet fuel. By changing the amount and type of fuel and the amount of oxygen or heat provided, it's possible to control how fast something burns. To put out a fire, one of the three things—heat, oxygen, or fuel—must be removed. In a wood fire, pouring water on it cools it off very quickly and stops the reaction from being self-propagating. This doesn't work with a grease fire, however, because oil floats on water. Throwing water on it causes the flaming oil to splatter making the fire more dangerous. Dumping baking soda on it starves it of oxygen and puts it out. It's not a great idea to use flour, though, as I'll explain shortly.

Figure 9.1 illustrates how the fire isn't burning the fuel but rather floating on top and igniting the vapors. If you are interested in making a cocktail that tastes like jet fuel and a hangover, this *Game of Thrones*–inspired "wildfire" cocktail was a combination of melon liqueur, orange juice, and watermelon vodka with a grain alcohol floater to make it light.

How Might Dragons Make Fire?

For a dragon to breathe fire, it needs three things: heat or sparks, fuel, and oxygen. The oxygen is easy to take care of if we assume the dragon has very large lungs. The sparks and fuel are a bit more difficult. Many have

FIGURE 9.1
Flames dancing atop a row of "wildfire" cocktails. The drink tasted terrible, but it's a good example of how fire burns the fumes rather than the fuel. Photo taken by Garrett Hamlin.

made suggestions, but here I'll address my favorites, or at least the ones I see as the most plausible.

Dragon fuel would need to be a substance that can be kept inside the body in large quantities and easily expelled, and it can't be too dense, seeing as dragons have to fly. One huge advantage for dragons is that many good fuel sources are organic. The most likely way a dragon could breathe fire is by igniting one of several organic fuels with a spark. The trick is finding a fuel that is both easily made and easily stored internally. Then, there needs to be some sort of ignition. I've read many theories on how dragons breathe fire, but the ignition mechanism always seems to be the one that is glossed over. Another rarely discussed issue is fire protection for the dragon. How is a dragon's mouth not as scorched as Dickon Tarly?

As far as a fuel source goes, there are a few reasonable options. The first one I'll talk about is based on my favorite demo, the birthday candle torch. (If you are daring, this is something you could try at home, but please don't sue me or my publisher if something goes wrong. I take no responsibility.) Normally, if you dropped a match on cornstarch, the match would just go out. That's because one of the three things needed to sustain a fire wasn't available: oxygen. Sure, there was some at the edges, but not near enough to keep a combustion reaction going. Cornstarch is an interesting substance in that its molecules are spherical. If you rub it between your hands, it feels soft and slippery. That's because it's a bunch of really tiny ball bearings. I could talk endlessly about the amazing properties of cornstarch (it's worth a quick Google, if you have a moment), but here I'll focus on its flammability. Like all starches, cornstarch is designed to be oxidized, one way or another. Starches store energy to be used by the body later. If you are an endurance athlete, you've probably "carb-loaded" to store up energy for a big race. Cornstarch is designed to store energy until it's needed. When it's around a flame, or some way of starting a combustion reaction, it releases that energy. If you blow air between the starch molecules and then expose them to a flame, it makes an amazing torch. Because the molecules are so small and spherical, each one has a lot of surface area exposed to the air, so one spark will make it go up in a really nice flame. This is why you occasionally hear reports of grain silos exploding—if there are a lot of grain particles in the air and they come into contact with a spark or other flame, it can cause an explosion. If you want to see this effect firsthand, you can fill a squeeze bottle with cornstarch, light a birthday candle, and blow some cornstarch over the flame. This takes a little practice to do without blowing out the candle and makes one heck of an interesting mess, but it also shoots fire the same way a dragon might. Magicians have also been known to do this trick, but they typically use something called lycopodium powder, which is slightly more flammable and much more expensive. Maybe a better idea would be to watch some nice, safe videos of someone else doing it on YouTube.

The one issue with this, though, is that if a dragon's "fuel bladder" were, say, punctured by a White Walker's Olympic-level javelin throw, it wouldn't explode. When Viserion's neck was punctured, there was a clear explosion out of the side of his neck. As much as I would love dragons to use some sort of powder as a fuel, that probably isn't the case. To cause an explosion like that, the fuel bladder would have to be full of some sort of pressurized gas. Methane is an example of one type of flammable gas. Cows are a big producer of methane. Microbes called *methanogens* in a cow's gut break up the cellulose in the grass it eats to release its nutrients, and methane is a byproduct of that process. If we assume that the average dragon is many, many times larger than a cow, with a correspondingly larger fuel bladder, this means they can produce a lot more methane. Methanogens can't function in an environment with oxygen. They are *anaerobic*. This is great news if methane is a dragon's fuel source. There's no way methane could be produced in an environment in which it could explode. The two processes, combustion and methane production, are mutually exclusive, so this seems like a great method for producing the fuel. The problem though, is that dragons don't eat grass—they eat meat. This may be great for the drama, but it's not so great for methane production. Because the anaerobic bacteria that create methane feast on the cellulose of plant matter, a meat-based diet is not going to produce the needed fuel. It very well may be that Balerion ate grass on the side but, seeing as it wasn't very "dragonly," no one ever mentioned it.

The question then becomes: Is it possible for a dragon to create enough methane to endanger all of Westeros just by eating a little grass on the side? A good way to estimate that potential is to look at cows' methane production. According to a study done on dairy cows, which produce about twice as much methane as beef cows, they produce about 20 g of methane per kilogram of dry food consumed. That's about 330 g of methane a day, or roughly 1 L of methane a day.[3] This is only about 0.04 cubic feet. Methane produces about 1,000 BTUs, or British thermal units, of heat per square foot when ignited.[4] A BTU is the amount of heat needed to raise the temperature of one pound of water by 1°F. In a single day, the methane produced from a single cow could raise the temperature of a pound of water by 40°F. That wouldn't even make room-temperature water boil. So, what if Drogon was the size of 100 cows with a similar digestive system? That would mean about 4 cubic feet of methane, 4,000 BTUs, and a few pounds of boiling water. This level of fire could definitely be dangerous; indeed, a cigarette can cause a forest fire, but it doesn't lead to the kind of destruction wrought by the Khaleesi's children.

For a cornstarch-like material or methane, a spark would be needed to start the flame. This is most likely the easiest part of creating dragon fire.

There are a few different options. One would be flint. Pterosaurs, which are potentially related to dragons, had gizzards, as do many birds. Gizzards help grind up food with the help of rocks. The bird (or pterosaur) swallows some rocks and gravel and they are stored in the gizzard. The muscular walls of the gizzard break down tough food with the help of stones. It's not too far of a stretch to assume dragons could have a "fire gizzard" that stores flint and iron pyrite. Usually steel is used with flint, but it would be hard for dragons to easily obtain steel. When flint is struck against iron pyrite, the reaction is very similar to the one created by touching steel wool to a 9V battery. When a tiny piece of iron is struck off with the hard flint, oxidation occurs quickly and creates a spark. A dragon could easily blow the contents of their fuel bladder over the spark to create a blowtorch.

The spark could also be caused by something like sodium, potassium, or another alkali metal. When alkali metals come into contact with water, electrons from the metal quickly transfer to the water. This creates ions that very strongly repel each other, causing what's called a *Coulomb explosion.*[5] In a Coulomb explosion, the ions repel each other so violently that they tear apart nearby molecules, releasing enough energy to cause an explosion (a demonstration favored by many chemistry teachers). It is entirely possible that dragons could house some sort of alkali metal, potentially in a different type of fire gizzard, and judiciously introduce water to cause a spark. Although this could work, it would be very dangerous. The alkali metal would have to be stored completely dry. This would be difficult to manage in an animal whose body most likely contains a large percentage of water.

A third ignition mechanism might be a piezoelectric crystal. Piezoelectric crystals create an electric spark when squeezed. The atoms in these crystals sit in a slightly asymmetrical way. When the crystal is squeezed, the atoms are jammed together, upsetting the usual electrical neutrality and creating an electric charge. In practicality, if a dragon had such a crystal surrounded by a muscle that could squeeze it at the right time, it could produce a spark as the fuel was expelled over it. Luckily for dragons, the collagen in bone has some piezoelectric properties.[6] Other naturally occurring materials can produce sparks, including quartz and even sugar.

Of the potential issues surrounding dragon fire, it seems that producing a spark is the least problematic.

There's one way dragons might have evolved to create fire that combines the spark part with the combustible material: *hypergolic fluids*. Having such fluids as a fuel would still lead to the type of explosion seen in Viserion's neck, would not require eating grass, and would be self-igniting. They seem like a good candidate. Hypergolic fuels are liquids that ignite when mixed together, no spark needed. They have been used in rocket propulsion, so they can definitely produce the flame needed for a dragon's blowtorch powers. Just as in other combustion reactions, there is a fuel and an oxidizer, but in this case, they don't need an external spark to start the self-sustaining reaction. In general, the chemicals used are toxic and corrosive, but this doesn't necessarily rule them out; hydrochloric acid is pretty darn dangerous, but stomachs don't seem to mind. Hypergolic fuels were first studied in the 1930s, and BMW developed a hypergolic engine fueled by nitric acid and various other compounds in 1940. The most common hypergolic fuels are hydrazine, monomethylhydrazine, and dimethylhydrazine, and nitrogen tetroxide is typically used as the oxidizer. But are these things a dragon's body could make and store? There's no particular reason why these substances couldn't be made; the bigger question is whether they could be stored in the body. All of these fuels are toxic and highly unstable, so if a dragon were to produce them, it would need something similar to a stomach lining but on a much larger scale. It's not entirely clear whether that is feasible; however, it's not out of the realm of possibility. The drawback is the weight. Dragons are already right on the edge of being able to fly—the added weight of that much liquid fuel would most likely render them grounded.

Let's say the dragon gets everything right and produces a jet of flame. There's still the question of protecting its throat from the heat. There are several methods that humans have designed to protect themselves from fire, such as Kevlar and fire-retardant gel, but none of those are naturally occurring. There are, however, several organisms that are able to survive in extreme conditions: thermophiles, hydrothermal worms, and the bombardier beetle. Thermophiles, a group of organisms that includes certain types of bacteria and other microorganisms, live up to their name—they love heat. In most organisms, excessive heat breaks down the enzymes required to sustain life, so they cannot survive, but thermophiles can survive temperatures as high as 284°F (140°C) because their enzymes continue to function at very high temperatures (figure 9.2). That's great, but the temperatures we're talking about are much higher than 284°F, as I'll explain in the next section; at temperatures as hot as dragon fire, the skin of known animals would be burned off completely. Similar to thermophiles, there are hydrothermal worms on the ocean floor that love both high pressure and heat; however, scientists recently discovered that they don't have enough insulation to survive in temperatures much higher than 140°F (60°C).[7] The bombardier beetle, by contrast, can tolerate high temperatures within its own body—it doesn't exactly breathe fire, but it can shoot boiling hot toxic chemicals at predators. When the beetle is threatened, it mixes two compounds in a vestibule. The lining of that vestibule acts as a catalyst, causing an exothermic reaction that produces

FIGURE 9.2
Microbial mats of thermophiles surrounding the water give the Grand Prismatic Spring in Yellowstone National Park its vivid coloration.

enough heat to bring the chemical mixture to a boil. The liquid is then violently expelled from the creature's rear end. The pressure from the expulsion mechanism closes off the openings in the vestibule, thereby protecting the beetle's organs from being melted by the boiling liquid. Dragons aren't so lucky, however; any fire created within the dragon's body would have to pass by crucial organs on its way to the dragon's mouth. And even if a dragon were somehow able to protect its organs, the bombardier beetle has the advantage of a hard, protective exoskeleton that can withstand the heat of the chemical reaction. Dragons have no such exoskeleton. So, although there are creatures that can protect themselves from intense heat, none of them come close to being able to protect themselves from fire hot enough to melt stone.

I wish more than anything that I could write a paragraph easily summarizing how dragons might be able to produce flame. That was my hope when I set out to write this chapter. As seen in the chapter on dragon flight, I am comfortable granting dragons a small, one-time exception to the laws of physics. Despite my best efforts, however, I would have to give dragons no small number of physics exceptions to make fire-breathing possible. Creatures exist that can handle extreme conditions, but nothing near what a dragon's throat would have to endure. There are ways to make biofuel sources that dragons could use to produce fire, but none of those would be light enough to allow dragons to fly, be produced in high enough quantities, or cause an explosion if the dragon's neck is punctured. The only thing dragons might be able to do is create a spark. I think this is one mythical occurrence that must stay mythical.

DIFFERENT COLORS, DIFFERENT SIZES

I know I just used science to burst your bubble on fire-breathing dragons, but let's assume I didn't—because dragons breathe fire, damn it! Let's grant the huge myth exception and say dragons can breathe fire. If that's the case and the laws of physics are being defied, are there laws that they do obey? How does their firepower compare to their size? We know that their necks can explode, and that the potential destruction is directly related to the heat and size of dragon fire. If the fire really is magic in some way and not caused by conventional physical and chemical methods, is it possible to look at the descriptions in the book and the depictions in the TV show to get an idea of how dragons grow both in size and firepower? One of the most dramatic scenes of destruction was Balerion's melting of Harrenhal. Not destruction—melting. After his attack, the impenetrable castle looked like dripping candles. I'd like to use this section to determine if Balerion could have done that and if Drogon, now that he is fully grown, might be able to follow suit. This requires first answering three questions: How hot were Balerion's flames? Is Drogon anywhere near his level? And would those flames have been hot enough to melt stone?

Balerion the Black Dread was so named not just for the color of his body but for the color if his flame. It was said that his flame was so hot it was black. From a physics point of view, that can't really be true. It's true that flames come in different colors, and I'll assume it's true that Balerion was black, but there is no such thing as a black flame. There are some pretty awesome chemistry demos that supposedly produce black "flame," but they are really just superheated vapors and not a true flame. Nevertheless, black flame remains a staple of the fantasy world. Though there is no such thing as black flame, flame color does have to do with something black: black-body radiation. In this case, "black body" refers not to Balerion but to anything opaque and nonreflective, such as a lightbulb filament or burning wood. Iron is often used as an example of a black body. When black bodies get really hot, they emit electromagnetic radiation. When hot enough, that radiation is in the form of visible light. It's then possible to determine how hot the flame is just by looking at the color of the light being emitted.

Black-body radiation has been observed for as long as such observations have been made. When forging weapons, blacksmiths noticed that the metal emitted different colors based on its temperature. At the turn of the 20th century, scientists began to try and explain the phenomenon. To understand black-body radiation, it's important to know two things: what heat does to electrons and how charged particles move. Temperature is a measure of how fast particles are moving around; the faster they move, the higher the temperature. If the particles were to stop moving completely, the temperature would be at absolute zero (0 K), which was discussed back in chapter 1. This is the point where there is no temperature at all. When charged particles such as electrons move, they emit electromagnetic radiation. The faster they move, the higher the energy of the radiation. They have to be moving pretty fast to radiate visible light. Lots of hot things radiate in the infrared spectrum, which is why heat-sensing cameras are able to see things our eyes can't—they can see energy radiating from objects at much lower levels. As the heat increases, the charged particles move faster and emit light at greater levels of energy. In the visible spectrum (the rainbow, ROY G BIV), the colors appear in order from lowest energy to highest energy. An object radiating yellow light is less hot than something radiating blue light. If something is heated to a high enough degree, it will emit light in all colors and therefore appear to be white. There are multiple wavelengths released at each temperature, as you can see in the graph (figure 9.3), but the peaks indicate the dominant wavelength. You can see that the location of the peak changes based on temperature.

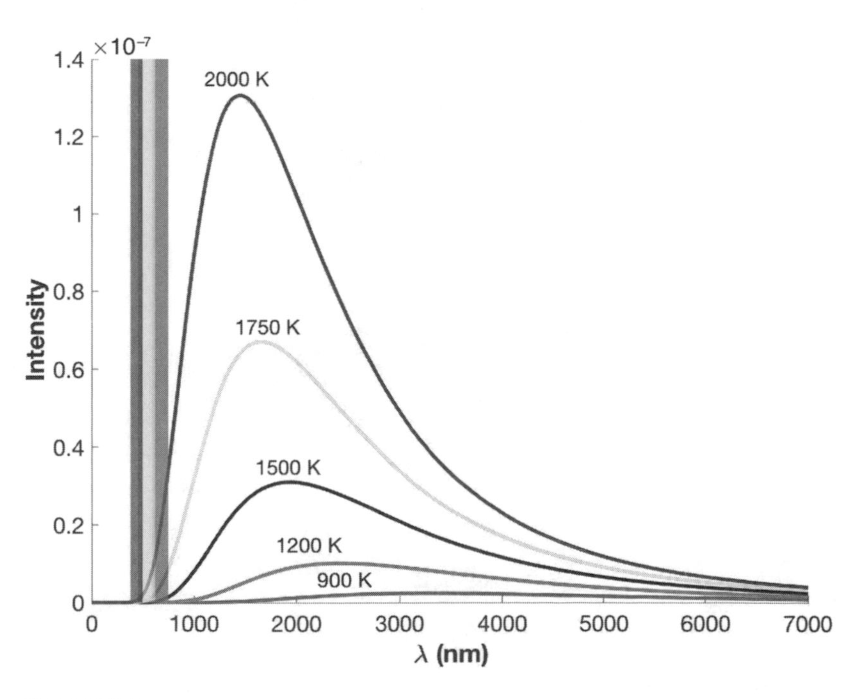

FIGURE 9.3
Black-body radiation spectrum at different temperatures. The gray area on the left side of the graph indicates the visible spectrum. Graph produced by Dr. Carolyn Kuranz at the University of Michigan.

It's easy to sum up this graph in a way that is actually useful: As soon as the tail of the graph of a certain temperature enters into the visible spectrum (390–700 nm), our eyes can see that color. When all the tails are in the visible spectrum, the glowing object looks white. The peak doesn't need to be in there, just the tail. The chart in figure 9.4 shows how the color of a fire indicates its temperature. In fact, you may remember from chapters 5 and 6 that smiths needed to know precise temperatures when forging swords. They could judge a blade's temperature by the color it was glowing. This chart was accurate enough to forge Damascus (or Valyrian, in this case) steel. It is not at all complete because it doesn't include blue fire, which is fairly common, but for dragon purposes it's a good start. Blue fire is hotter even than white. When the tail of the graph in the blue

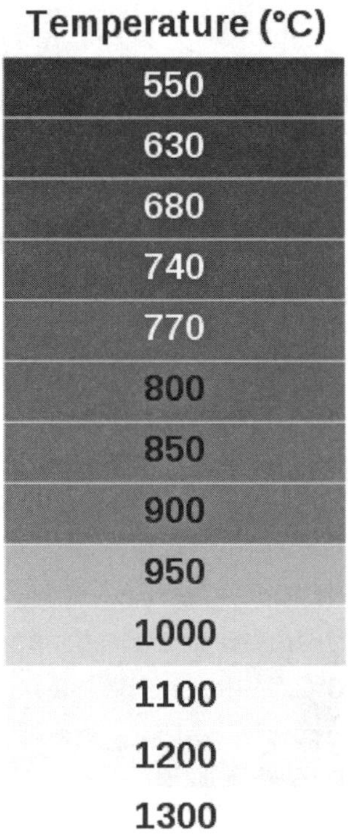

Temperature (°C)
550
630
680
740
770
800
850
900
950
1000
1100
1200
1300

FIGURE 9.4
How temperature compares to color of black-body radiation. This is handy for both blacksmiths and dragon victims. Even though this image is printed in black and white, you can see how the fire gets lighter and whiter as the temperature increases.

section of visible light is sufficiently higher than the other colors, the flame goes from looking white to looking blue. It is possible that when Westerosi oral historians referred to Balerion's fire as "black," they really meant "blue."

The story doesn't fully end there, but it's a good start. Fire color isn't only based on black-body radiation. When different chemical elements are added to a fire, it's possible to produce different flame colors based on how the elements behave when they get hot. When certain compounds are heated, the electrons jump up in energy levels. When they drop back down again, they emit light. The color of the light emitted depends on the difference between the high and low energy levels. Because the interval between the high and low energy levels of each element or compound varies, the colors are different. If, for example, you throw sodium on a flame, the flame turns yellow, whereas copper gives the flame a greenish hue. Chemists use the aptly named "flame test" to discover the composition of an unknown substance, and certain birthday candles rely on the same effect to create colorful flames. (If you want to try this at home, you can find instructions online for making pine cones that will add some color to the flames in your fireplace—it makes for a great new holiday tradition.) In all of these cases, the color of the flame changes depending on how the compounds' electrons are moving between energy levels, not the temperature of the flame. For dragons, however, I think it's safe to focus on black-body radiation instead of elemental influence on flame color.

All this is well and good, but what does it mean for dragons? It means that by looking carefully at the dragon fire in the show we can tell how hot it is. In season 3, Drogon made his first real kill when he torched Kraznys mo Nakloz with his yellowish-orange flames after the slave trader handed over the Unsullied in exchange for the dragon. That puts his flame at roughly 900°C. When he reappears in the fighting pits of Meereen at the end of season 5 and kills again, his fire is much lighter in color. This indicates a temperature closer to 1200°C, far hotter than before. It would seem that as Drogon grows in size, the temperature of his fire increases. Personally, I always wondered why Dany waited so long to use her dragons in Westeros—they're flying flamethrowers! That had to be worth something. But it seems her little blowtorches just weren't hot enough yet. Balerion's fire was, if you look at fan art, a dazzling white, or if you read the books, black, which I think we can assume is blue. This puts the temperature at 1800°C or higher. Would that have been enough for the Black Dread to have melted Harrenhal? If so, is Drogon now big enough to follow suit and take on the Red Keep?

What Is Melting, and Can It Happen to Stone?

Clearly, dragon fire is pretty hot, but the next step in understanding the destruction it causes is looking at how stone melts and determining the temperature required for that to happen. At this point, some of you are probably thinking it's impossible to melt stone—only things like ice and metal can melt. But stone can melt, too. If you've ever been lucky enough to visit Hawaii, you may have seen melted stone—I'm talking specifically about lava. Chapter 2 focused on ice, from how it's structured to how it can build a structure. One of the important takeaways from that chapter is how ice melts; in particular, how it melts under pressure. As we all know, though, ice—and everything else, for that matter—melts when it gets hot, too.

The castles of Westeros are made of granite (according to the descriptions in the book, at least). Granite is an igneous rock, meaning it was created by magma solidifying under the heat and pressure of Earth's mantle. Granite is typically a combination of quartz, feldspar, and other trace compounds, though the percentages of each mineral can vary widely. It comes in several different colors, including the pink of the Red Keep. Granite is formed by solidifying molten rock, so it makes sense that extreme heat can put it right back into a liquid state. Granite is extremely strong as a building material, and that strength comes from the way the crystals form. It's certainly not the only rock with that particular chemical composition, but because the rock cools slowly deep underground, it has time to form complex crystals. As the rock cools, the different minerals begin to solidify into crystals at different temperatures and different rates. Granite isn't composed of one single type of crystal the way ice is, but rather from different crystals interlocking. In chapter 2, I talked about how ice cracks along fracture planes and that introducing something like sawdust strengthens the ice by putting little roadblocks in the fracture planes. Granite is strong for the same reason: the crystals interlock, so there aren't really defined fracture planes. The fracture plane of one crystal has another crystal as a roadblock. You can even see these interlocking crystals with your eye; its speckled, grainy appearance is what gives granite its strength as well as its name.

Because the crystal structure and composition of granite differs from those of ice, it means that granite melts differently, too. In ice, there is one specific melting temperature, and when that temperature is reached and energy is continuously added to the ice, the solid turns to liquid. The extra energy needed to melt something, as I mentioned in chapter 2 in relation to melting ice, is called the *latent heat of fusion,* and it's about the same for granite as it is for ice. Although the basic principle is the same—energy must be added to raise the material's temperature to the melting point, at which point additional energy (the latent heat of fusion) is needed to fully melt it—it works a little differently with granite because granite's crystals have different melting temperatures. The majority of granite is composed of feldspar interlocked with quartz, the first and second most abundant compounds on Earth, respectively. Quartz has a melting temperature of roughly 1650°C, whereas feldspar's melting point is 450° lower at 1200°C. That, however, is not the entire story. The two minerals have different melting points, so you've probably guessed that granite is very different from ice in that granite can be mostly melted and yet still slightly solid. It's possible for the feldspar to melt but still have pieces of solid quartz within the flowing molten rock. To make it even more complicated, quartz doesn't just up and melt; the molecules twist, crack, and then melt. In the same way that ice comes in a number of different crystal structures, quartz and feldspar also come in different structures. Furthermore, quartz is structured in such a way that it doesn't simply melt at a given temperature; rather, it heats up until its structure changes, cracks a bit in the process, and then decides to melt. Normal quartz has atoms of silicon and oxygen arranged in a slightly twisted pattern. This twisting pattern makes it a *chiral* molecule, and the direction in which it twists is called the *chirality*. When quartz hits 1063°C, the chirality abruptly changes directions. This is a pretty jarring process for the crystal and usually leads to some cracking. Feldspar is much less interesting. Add heat and it melts. Putting this all together, here's what happens when granite gets hot: Not much happens until it gets to 1200°C, at which point the feldspar starts to melt and the quartz starts to crack. The stone will start to flow a little bit, but it's not fully melted; there will still be chunks of quartz within the flow of molten feldspar, cracking as they go along. As more heat is added, the quartz begins to melt, and after reaching a temperature of at least 1650°C, the stone will be completely molten. For Balerion to have given Harrenhal its characteristic melted candle appearance, his fire would need to have reached at least 1200°C, but getting it all the way to 1650°C would result in a more uniform melted effect. For Drogon to melt the Red Keep, then, he'll need to blast the granite with a sustained source of heat of at least 1650°C. Would that be possible?

WHAT ABOUT HARRENHAL AND BALERION THE BLACK DREAD?

This is the point in the chapter where I go through all the physics I've explained so far and tell you how science could make what seems fantastical possible. I've done that for most other chapters; in this case, however, there are a lot of obstacles to overcome for this to work. Although many have made vague arguments about how dragons might produce fire, when one looks at the math and science behind it, it just isn't possible. It might be possible for animals to breathe fire on a small scale—like, beetle-small—but for something the size of a 747 to produce a jet of fire worthy of an industrial flamethrower is beyond the laws of physics. Dragons might be able to fly, but flame is just a burned bridge too far.

Let's take the completely unreasonable leap in logic anyway and assume that, thanks to a healthy dose of magic, dragons can be the flaming beasts they are described as. The science question now becomes whether or not Balerion—or, more importantly, Drogon—could melt a castle. We'll start with Balerion. Although there are no official pictures of Balerion, the books' description puts his fire in the bluish range, or about 1800°C, as I said earlier. This is certainly hot enough to melt the feldspar in granite and just enough to melt the quartz, too. The trick, however, is with the time it would take. Latent heat of fusion makes everything more complicated. Either Balerion or Drogon could crack Harrenhal and leave it in ruins, but it would take time for either of them to make the castle melt. As I said in chapter 2, it would take about 5.5 minutes for the tiny coils of a red-hot space heater to fully melt a kilogram of ice that had reached its melting temperature. A dragon has the benefit of higher flame temperatures as well as the ability to cover a wider area with its flames. Assuming Balerion spouts a square meter of white-hot flame, he could melt that same kilogram of ice in about a minute and a half, which isn't that much time. But granite, as always, is more complicated. First, there's the time it would take to raise the blocks of granite to the melting temperature. From there, it's another minute and a half of direct heat to melt the stone—and that's just for a kilogram of stone. A kilogram of ice is about 10 cm^3, but a kilogram of granite is much, much smaller. I'm guessing the turrets of Harrenhal were built from a whole lot more than a kilogram or two of granite. Much like Balerion, Drogon was big enough by the end of season 6 to produce fire hot enough to melt feldspar and almost hot enough to melt quartz at a sustained flame. Considering his flames grew hotter as he grew bigger, his flame will soon be hot enough to melt stone. If we gloss over the fact that there's no plausible way for an animal to make fire on that scale, Balerion and Drogon would certainly be capable of producing the kind of castle-melting heat that puts Targaryens on the Iron Throne. The question, really, is whether he has the patience for it. And, of course, just when things start to make sense, GRRM up and kills a character, leading us to both soul-crushing sadness and a whole new branch of science.

VISERION'S MAGIC FIRE

Run-of-the-mill yellowish-white fire was just not enough for the huge world of *Game of Thrones*. At the end of season 7, there was some pretty impressive blue "fire." As Neil DeGrasse Tyson correctly stated in a tweet, and as I've described elsewhere in this chapter, blue fire is much hotter than orange fire.[8] However, there has been much discussion as to whether it is normal blue fire or some sort of freezing anti-fire. Just such a flame was first described in GRRM's book *The Ice Dragon*, which was both shockingly short and written for kids. I didn't think he could write a book that accomplished both those things. Fandom seems split on whether or not Viserion's fire is very hot or, as GRRM described the breath of an ice dragon in *A World of Ice and Fire*, "a chill so terrible that it can freeze a man solid in half a heartbeat." The Wall came down due to the blasts from Viserion, so it seems clear that it must be hot flame. But whether the flame is hot or cold, it's still not that simple. A process called *thermal fracture* happens when there are temperature gradients within a material. If you've ever seen a glass window fracture when it's warm inside and cold outside, you'll know what I'm talking about; it happened to my windshield once. It's possible that if Viserion's breath were like an ice dragon's, he could blow enough cold air to knock the Wall down via thermal fracture. As I've already said, however, melting by adding heat isn't a straightforward process either. It takes heat to raise the temperature to the melting point plus even more heat to actually melt something. Could blue flame even accomplish that? The action seen at the end of season 7 might happen with hot dragon flame or ice fire, just through different physics . . . or maybe it can't happen at all. Let's look at both options and see if one, both, or neither makes more sense given how it was presented visually. It would be much easier if GRRM would just get to writing about this bit already so I had some actual text to go off of, but alas, even if a third or fourth edition of *Fire, Ice, and Physics* is released, I doubt I will have much more source material to work with by then.

It's been so much fun talking about melting in this chapter, so let's keep rolling with that. I explained how stone melts and how the latent heat of fusion is what actually makes it melt. The same is true of the Wall. Viserion's fire would have to be both extremely hot and capable of expending energy quickly enough to melt the ice. Blue fire is hotter than white fire and produces about twice as much energy per square meter. This means it would take about 45 seconds to melt a kilogram of ice. That seems pretty darn quick, but remember, that's one 10 cm cube of ice out of an extremely thick wall. It would take Viserion about 45 seconds to make even a very small nick, maybe the size of a wildling's crampon, in the giant Wall. Even though blue fire has about twice the energy per square meter of white fire, it's still just not enough to bring down a wall, even a questionably built one. It's even worse when you watch the episode and see him cruising along the length of the wall at a fast clip. I hate to say it because it seemed so likely, but no dragon could take down a wall like that, zombie or otherwise.

Maybe thermal cracking can take this from magic to reality? Before going into the specifics for ice walls, here's a general overview of why things crack when one part is hot and one part is cold: For something to crack, it needs to encounter a force of some kind. In most solids, the molecules spread apart when heat is applied. I talk a lot about what happens to glasses when they get cold in chapter 8, but that discussion didn't involve parts of the material getting hot and others staying cold. When the temperature in one part of a material differs from the temperature in another part, it creates stress. The molecules on the hot side speed up and spread out while the ones on the colder side either slow down or condense. This creates stress, and, in many cases, if the temperature difference is extreme enough, the material will crack, sometimes explosively. This is of particular concern to laser scientists who are dealing with crystals that get very hot and could crack if not cooled well (pew pew pew!). When Viserion hits the Wall with a very cold flame, he is cooling down one part of the Wall while the other stays warm. This temperature gradient causes stress, so the question becomes: Is it enough stress to make the Wall crack? Add to that the fact that parts of the Wall might not be fully frozen.

If you've read chapter 2, you know that ice is very, very weird. Pretty much anything that can be said about normal solids doesn't apply to ice. Ice does, however, lose density, or contract, as it gets colder. If it's hit with an icy blast, there will be stress from the ice contracting as it cools. When it's cold, it's also stronger. One group found that ice actually gets significantly stronger the colder it gets. The force needed to crack ice just from whacking on it increases as temperature decreases. If zombie Viserion was blowing icy fire, he might actually have been strengthening the wall.[9] But what about the stress caused as ice gets colder? At liquid nitrogen temperature (–196°C), there is enough stress between the outside and inside of an ice cube to make it crack pretty quickly. This doesn't exactly scale up well. Sure, if ice flame is the temperature of liquid nitrogen, it will definitely crack the outside of the Wall, as seen in the show. But, again, we run into the pesky problem of the Wall's thickness. There is simply no way to crack the Wall all the way through with the cold.

As I said in chapter 2, the Wall is most likely reinforced with sawdust and a prayer, so I don't think it would take much to knock it down, but the power from Viserion, whether hot or cold, would not be enough to cause the wall to crumble. We can now all rest easier knowing that science says the Seven Kingdoms are safe, and Tormund will be just fine. It looks like Martin should have gone with the fan theories of freezing the Shivering Sea at Eastwatch and allowing White Walkers to just walk across. Now that is something Viserion the ice dragon could have pulled off.

10

The Battle of the Blackwater

The Science of Wildfire

The kiss of wildfire turned proud ships into funeral pyres and men into living torches.

—Tyrion Lannister, *A Clash of Kings*

Wildfire is probably one of the most destructive weapons—next to dragons, of course—in Westeros. It's explosive, it sticks to things, it can burn on water, and it's green. It is also said to melt wood, stone, and steel. Water can't put it out and it gets more potent with age. This is some pretty scary stuff. So, is it real? We have already seen that fire can be hot enough to melt stone and steel, and that water shouldn't be used to put out an oil fire. Is it possible to have fire so hot that water turns to steam before it hits the fire? Most things, except for maybe wine and cheese, lose potency with age, but wildfire strengthens over time. What types of compounds might behave similarly? As in the other chapters in this book, it's probably best to start by examining the fictional claim's real-world effects and then seeing if there is a way to create a substance that can replicate said effects. It seems like that would be a good thing to do here, too. I explained how fire works in the last chapter, so we don't need to go through that again, but wildfire is uniquely fearsome in contrast to plain old dragon fire.

As any fan of the show clearly knows, GRRM is an incredible amateur historian. Much of his writing draws heavily from historical records, and quite accurately as well: the Targaryens' inbreeding, Valyrian (Damascus) steel, and even the Red Wedding are all rooted in history. Wildfire is no exception. The Greeks were way ahead of GRRM, having developed one of the first chemical weapons, Greek fire, in the 7th century. As with the secrets to making Damascus steel, the specific composition of Greek fire has been lost to the ages; however, information about its potency and hints about its components still exist. It's possible that the legends surrounding its destructive power have been exaggerated over time, as tends to happen, but by all accounts, this substance was frighteningly destructive. Wildfire is most likely a green version of Greek fire, and I promise that I will discuss it in detail. If you've read this far, however, you know I'm nothing if not thorough, so I'm going lay out some background information first. Perhaps the most interesting aspect of wildfire is how it interacts with water, so I'll start there.

WATER DOESN'T ALWAYS BEAT FIRE

In chapter 9, I talked about how water puts out fire. When water is thrown on a fire, it cools off whatever's burning enough to stop the self-propagating combustion reaction. Grease fires, however, are different. If you've ever lived in Texas, you've probably deep fried a turkey (I certainly have, at least), and hopefully you read the dire warnings in the deep fryer's instruction manual. If so, you know how dangerous frying a turkey can be, particularly if you try to do it in your house. (Still, the delicious end result is well worth the danger.) That's because grease fires are Class B fires, which explode when they come into contact with water (so you should always dry off your turkey before slowly lowering it into the oil). Fires are classified by what is needed to put them out, so Class B fires obviously can't be put out with water. When water hits a Class B fire, or when a wet turkey drops into a vat of hot oil, the water turns to steam almost immediately. When water turns to steam, it expands, causing the fire to explode and fling scalding oil in every direction. Most oil-based fuels such as petroleum, tar, grease, and lighter fluid burn as Class B fires. One thing to note about most of these fuels is that they also float on water. A fuel that burns hot, floats on water, and bursts into flame when it gets wet . . . sound familiar?

You can start one heck of a fire in your kitchen if oil floating on water in a frying pan catches fire, because the water underneath the oil won't stop it from burning. Now, let's apply this logic to an oil slick floating down a river. If a body of water is polluted enough, it can, indeed, catch on fire. The Cuyahoga River, which runs through Cleveland, Ohio, proved this repeatedly, having caught on fire not once, not twice, but an estimated 13 times, the first of which happened in the 1860s and the last of which occurred in 1969. It took the environmental embarrassment of a burning river to launch the Environmental Protection Agency (EPA), by the way. Back in the day, rivers were basically dumping grounds for just about anything: sewage, oil refinery waste, chemical waste, debris—all of it ended up in rivers. Given the Cuyahoga's proximity to the thriving city of Cleveland and its many factories and oil refineries, a lot of petroleum waste ended up in the river, to the extent that *Time* magazine once said that the Cuyahoga "oozes rather than flows."[1] J. D. Rockefeller's Standard Oil of Ohio, one of the largest refineries in the history of the United States, was a huge contributor to the pollution because it regularly disposed of waste products—one of which was gasoline—in the river. You can probably see why this was a bad idea. Oil and other petroleum-based pollutants float on water and are flammable. The pollution in the river became so thick that it is said to have formed a three- to six-inch layer on top of the water. The worst of the fires occurred on November 2, 1952, and caused an estimated $1.5 million ($14.2 million when adjusted for inflation) in damage. Because the thick layer of oil created a barrier between the river's water and the fire, water couldn't break through to extinguish the flames. The last in the long string of river fires occurred in 1969 and triggered the creation of Earth Day and the establishment of the Clean Water Act and the EPA. And all it took was a photo of a burning river in *Time* magazine.

THE DANGERS OF POLLUTION

You may be wondering how a grease fire of that scale is extinguished; after all, there's only so much baking soda one can dump on a giant burning river. Accounts of the 1969 fire report that it only took firefighters about 20 minutes to contain the blaze. That fire, however, started on a small oil slick that floated under a bridge, which also caught on fire. Firefighters used water to put out the flames on the bridge as well as the oil slick. The containment of the fire was probably aided by the fact there was only a limited amount of fuel and that the oil slick was isolated; it's not like the fire could have jumped to anything else. There isn't much documentation on how firefighters were able to control the flames in the 1952 fire, but they probably took a similar tack: isolating the oil slick, using water to calm the flames from debris, and letting it burn itself out. In fact, strategic burning is one way still used to "clean up" oil spills. Unlike the Cuyahoga River's sludge, however, most oil spills that occur in the Gulf of Mexico or elsewhere are not that thick, so there's some extra work required to clean them up. It takes a lot of effort to successfully burn oil on water, but it's easier if the oil is collected into a slick thick enough to burn. If there is little wind and the sea is calm, the slick is ignited and the fire controlled until the oil burns off and it goes out. The US military works darn hard to do what those in Ohio were able to do by pure luck.

We've established that fire can burn on water on a grand scale as long as the fuel is less dense than water. So, to weaponize this, you'd need something petroleum-based, thick enough to combust, and capable of burning for a long time. Pouring gasoline or crude oil onto water and dropping a match wouldn't be enough to catch all of Blackwater Bay on fire, so wildfire must also be thicker than oil alone. The good news is, though, that at some point it will probably just burn out on its own. The military buffs and the few people who have studied wars more recent than WWII in high school history reading this probably know where I'm going next: napalm. Napalm has a well-deserved horrible reputation, and it is a troubling subject to discuss even in the abstract. Unlike the carnage in *Game of Thrones*, the horrors of napalm were very real and may have directly affected you or your loved ones. I will do my best to talk about the science of it as it pertains to wildfire, but if you don't want to read about it, skip to the next section, "Colored Fire."

MODERN FIRE WEAPONS "PERFECTED"

Napalm has become a generic term for fuel, usually gasoline, mixed with a thickening agent. There are several different variants of the substance and the term is general enough that there's no single chemical formula to track down. It's characterized by being both sticky and slow burning. It can burn at temperatures of 2200°F, can't be washed off, and some formulations can burn while fully submerged in water. It is not stored in its gel form but rather as a powder that is then mixed with gasoline when needed. The name comes from the two compounds that were first combined to form the powder: *na*phthenic acid from crude oil and *palm*itic acid from coconut oil. Note that these are substances that the alchemists of Westeros would have been able to create. What differentiates one variety of napalm from another is the type of thickening agent used. This also changes how the gel sticks, how long it burns, and how hard it is to extinguish. Napalm B, one of the more destructive forms of napalm, is made with polystyrene (Styrofoam, basically) mixed with gasoline.

Napalm, invented in 1942 by Louis Fieser and E. B. Hershberg, was first tested on the Harvard College soccer field.[2] Both inventors initially saw it as a way to burn plant life and buildings in wartime and never imagined it would be used on people. Though horrified by its eventual use on humans, both have stated that they take no responsibility for napalm atrocities. Interestingly, Fieser went on to have a successful career developing blood-clotting factors, synthesizing vitamin K, and developing antimalarial drugs. His official biography from the National Academy of Sciences makes no mention of napalm. It is also extremely hard to find information on the chemistry of napalm except through infamous sources such as *The Anarchist Cookbook*. After the internet searches I have done for this book, I have undoubtedly earned my own FBI file. I hope Scully is impressed.

Luckily, the military historian friend (we'll call him Ted) I mentioned when talking about bergships in chapter 2 is excellent at finding hidden sources. The term "napalm" is much like the term "Kleenex" in that it has become a catch-all term for any gelled explosive. The military term for it is "incendigel," for obvious reasons. The original compound involved mixing naphthalene or palmitic acid with gasoline to thicken it up and to slow its burning time. As discussed in chapter 9, it's the fumes that burn, not the fuel itself. The fuel needs to be in a gaseous form for combustion to take place. Normal petroleum products are as combustible as they are because so many fumes are present even at room temperature. In the gel form, fumes are not readily available for combustion. This makes napalm stable and long-burning. In fact, it was stable enough that the army would ship it in 165-gallon containers. Like wildfire, it will remain potent for a very long time, although it does not become more potent with age. Staying potent at all over a long span of time is still pretty impressive. Traditional napalm would burn for roughly 30 seconds. Some sources claim that it's possible to control the burn time based on the mix of petrol and thickening agent. This may very well be true, but I haven't been able to confirm it with primary sources. In addition to longer burn times, thickening agents allowed for more effective delivery mechanisms. Napalm was used in various types of bombs, land mines, and flamethrowers. The latter became significantly more effective with the thicker substance, which increased the flamethrowers' effective range fivefold. A single firebomb dropped from a low-flying plane could have a blast radius up to 2,500 yd^2 in size. This stuff was no joke and was infinitely deadlier than gasoline alone. Keep in mind that in these areas of destruction, it wasn't just fire but sticky petroleum-based fire that couldn't be put out easily—or with water, in particular.

In the 1960s, Dow Chemical took standard napalm and kicked it up a notch by creating napalm B. When you think of the iconic images of napalm attacks, you are thinking of napalm B. Burn times increased from 30 seconds to 10 minutes or more. It was deployed almost exclusively as an antipersonnel weapon, which is a fancy way of saying it was used to burn people rather than strategic targets. Napalm B's thickening agent was changed to polystyrene, and some benzene was added to create what's now most commonly known simply as "napalm." The properties that made napalm so deadly in the first place are its petroleum base, which causes class B fires; its thick, sticky texture; and the fact that it burns slowly instead of causing one big explosion. Polystyrene and benzene only enhanced these lethal characteristics. Polystyrene is not very flammable—it actually acts as a bit of a fire retardant—but it thickens when dissolved in gasoline and acts as a buffer to the combustion reaction. Benzene burns more slowly than gasoline, extending the burn time of napalm B. Replacing naphthalene or palmitate with benzene and polystyrene resulted in a 20-fold increase in burn time. The longer burn time meant the substance was less likely to explode and therefore easier to handle.[3] Napalm's stability was key to its "success" as a weapon. It's not much good to have a weapon that is as likely to burn you as it is the enemy. In fact, it can be quite difficult to ignite napalm. Designing the delivery system was almost as important as designing the fuel. Current napalm B bombs are encased in thin aluminum shells that shatter on impact when dropped from a plane. Originally, napalm was lit with thermite, but this was not a reliable method because thermite itself is notoriously difficult to ignite. Now, napalm is most often lit with white phosphorus injected into the gel upon impact. White phosphorus (WP, or "Willie Pete" in military slang) is no picnic on its own—it causes severe burns and can melt skin—but it lacks the burn time of napalm.[4] It is self-igniting and will begin a combustion reaction when exposed to air because its ignition temperature is approximately 86°F (30°C), but it can reach temperatures as high as 5000°F (2760°C) while it burns. The reaction also produces a lot of smoke, so WP was often used for tracking the trajectory of the bomb because the volume of smoke made it easy to follow. This makes it even more deadly when used near people—if they aren't burned, they are injured by smoke inhalation. When the firebomb hits its target, the aluminum shell will break on impact, scattering the napalm up to 180 m (or about two football fields' worth of destruction). The WP is placed in such a way that as soon as the aluminum shell shatters it is ejected into the scattering napalm, which ignites as soon as it is exposed to air. It is frighteningly efficient.

It sounds like napalm might be a good candidate for wildfire, and that's not untrue. It certainly burns hot enough and can definitely burn on water. It could melt stone and steel and is almost impossible to put out. There are three main issues with this, however. The first is that it isn't green, and wildfire is clearly green. In the next section I'll talk about ways to maybe get around that, but fundamentally, napalm is not the dramatic green color needed. The second issue is that it is very, very stable. This is one of the reasons it has been so prized by the military. It is safe to ship, easy to handle, and getting it to actually ignite is tricky. There is no way Bronn would have been able to ignite the whole of Blackwater Bay using a single flaming arrow if the fuel were napalm-like. Once napalm is burning, however, it is as deadly and as feared as wildfire. The third issue is that the most destructive form is napalm B. Although original napalm is quite destructive, it does not live up to its fictional counterpart. The main drawback is that it is easier to put out and it has a far shorter burn time than napalm B. It would make sense, then, to say wildfire is napalm B, but it's not that easy. Second-generation napalm is made with polystyrene, which is not found in nature. We encounter it every day, but I don't think Westerosi couriers were seeing many parcels stuffed with packing peanuts. This isn't a deal breaker, however. The very first polystyrene was synthesized in 1839. That's not exactly as primitive as Westeros, but it's not completely modern, either. A German apothecary named Eduard Simon distilled a resin from the sweetgum tree and let it sit, and came back later to find it had thickened into a jelly. After several more chemists experimented with similar products and processes, the final compound was dubbed polystyrene. In 1941, Dow Chemical invented the Styrofoam process and made it into the ubiquitous product we know today. So, there is no inherent reason that the alchemists of Westeros would not have access to a compound similar to polystyrene. If they did, they certainly could have made it into a napalm B–like wildfire compound—but it still wouldn't be green.

Colored Fire

One of the most distinctive aspects of wildfire—both the substance itself and the flames it produces—is its green color. How could the fire be green? If wildfire is similar to napalm or Greek fire, it burns at around 1200°C. A fire that hot should be clear orange to white. If you've seen pictures of burning napalm, you know it's not green—but could something be added to such a mixture to make it burn green? One aspect of colored fire that I handwaved in chapter 9 was how fire could change color through means other than temperature. I'll fix that oversight now.

It's not hard to find something that will produce green fire; these days, everything from novelty birthday candles to festive pine cones can be made to burn in an array of colors. The science behind how these colors are produced is different from how a flame changes color at different temperatures. Instead of the color coming from black-body radiation, it's a side effect of what happens to different elements when they're exposed to heat. The range of colors produced is called the *emission spectrum*, and it is unique to each element. The structure of the element's atoms causes the color and it doesn't matter how the element gets hot, whether it's from an electrical charge or exposure to a flame, but when it gets hot it emits light in a very specific way. Hold onto your hats, because this explanation is going to involve some quantum mechanics.

Think back to that picture of an atom you were shown back in elementary school. You know, the one with the nucleus in the middle and the rings of electrons orbiting around it. It's called the Bohr model, and it is simplistic and entirely incorrect; however, it is the best model for discussing an element's emission spectrum (figure 10.1).

In figure 10.1, you can see the different rings around the nucleus. Electrons have preferred rings, or *orbitals*, they like to hang out on. When each of an atom's electrons is sitting on its preferred orbital, it's called the *ground state* of the atom. But electrons can jump around, jump around, jump up, jump up, and get down—literally. Quantum mechanics says these energy levels are quantized and that electrons must be on one orbital or another; they can't float around in the middle. When enough energy is added to the atom, whether it's through electricity, or heat, or something else, electrons jump up to a higher energy level. They can't slide up or move gradually toward the next energy level; they have to jump all the way up. There is a specific energy difference between each orbital in an atom. For an electron to jump up to an orbital with a higher

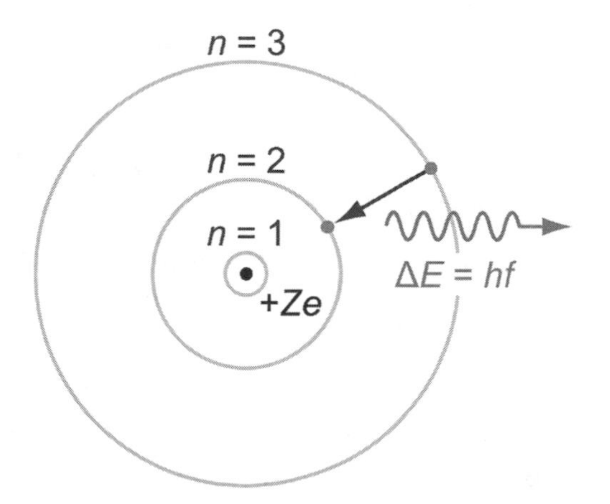

FIGURE 10.1
The Bohr model of the atom. When an electron falls down from a higher energy level to a lower one, it gives off light. The color of the light depends on how far it is falling.

energy, it has to be given at least that much energy. (Think of it like the "you must be this tall to ride" signs you see at amusement parks, but for energy instead of height.) The electron hangs out at this new energy level for a while and then gets down (we usually say "falls down," but those aren't the lyrics, now, are they?) to its ground state either by going right back to where it started or hopping through other lower energy levels on its way down. When it falls down, the energy it lost has to go somewhere. It's emitted as light (the red squiggle on the figure). The color of light emitted depends on how much energy is lost. Different energy transitions give off different colors of light. Lower energy transitions are redder, higher energy transitions are bluer. When energy is added to an atom, several transitions occur and the atom may emit more than one color of light. Those colors will be the same every time, however—remember, atoms can't be in between energy levels and the difference between each energy level is specific to each element.

Once energy has been added to an atom, things aren't as straightforward as the last paragraph may lead you to believe. Each element has different energy levels in its atoms and the energy levels aren't all the same distance apart. It might take a different amount of energy to jump from the first orbital to the second than from the second one up to a third. To further complicate things, electrons don't have to jump to the nearest orbital—they can jump a lot higher depending on how much energy is added. Then, when they fall back down, they might fall in different stages—think of it like releasing a ball from the top of a staircase and letting it bounce its way back down instead of throwing it straight down to the bottom. All of these transitions are different for each element, which is why each element emits a characteristic spectrum light when energy is added. Neon lighting is a great example of this effect. A neon sign is made of tubes filled with a number of gases (the original being neon) that produce different colors when electricity is passed through them. What you see when you look at a neon light is a mixture of the colors produced by the electrons jumping around. The same thing happens when elements are burned, so you can make a guess at what elements were added to a fire just by looking at the color of the flame. This method is often used in labs to puzzle out the composition of an unknown substance. Using the fittingly named flame test, researchers can burn a sample, note the color of the fire, and make a darn good guess at the elements within the sample.

So far, we've established that wildfire is probably something similar to napalm, but that doesn't explain its distinctive color. Boric acid or copper sulfate, however, might. We know it's not the temperature that makes wildfire green, but the presence of a chemical element in the mixture could explain its signature color. There are roughly 10 compounds that will make a fire burn green, but copper sulfate and boric acid are the most vivid. There are plenty of instructions online that will tell you how to make green fire safely at home, and I would highly recommend trying some of them out if you ever decide to throw a *Game of Thrones*–themed party.

One particular boron compound is getting a lot of love on the internet for being the real-life wildfire, and there's a strong case for it. Trimethyl borate burns with a beautiful green flame. It is also quite difficult to synthesize without injury and can be explosive. It can't be extinguished by water, but it also can't float on water, either, and that's one of the hallmarks of wildfire. Still, its hazard summary from the New Jersey Department of Health and Senior Services is pretty impressive:

> Trimethyl Borate can affect you when breathed in and may be absorbed through the skin. Contact can irritate the skin and eyes. Breathing Trimethyl Borate can irritate the nose, throat, and lungs. High exposure can cause headache, nausea, vomiting, diarrhea, loss of appetite and convulsions. It may damage the kidneys. Trimethyl Borate is a FLAMMABLE LIQUID and a DANGEROUS FIRE HAZARD.

Sounds fun to me! It is also synthesized from things found in nature, so it is something the alchemists could have created with limited knowledge of modern chemistry. Of all the boron compounds, this is the most likely real-life wildfire.[5] That said, it doesn't burn on water, it's colorless in liquid form, it burns at 816°C—not nearly hot enough to melt any of the requisite items, and it doesn't burn for very long, either—so we still haven't found the magic compound.

In the discussion of dragon fire, I talked about hypergolic fuels. These are liquids that don't need any type of flame for ignition; they catch fire simply by mixing. It's possible that wildfire could be of this type. It would certainly explain how carefully it must be handled and its explosive power. Diethylzinc is a highly reactive hypergolic liquid that is often used in rocket fuel. It reacts violently with water and ignites on contact with air. Oddly enough, the US Library of Congress once used it in an attempt to deacidify the wood pulp paper found in some older books in its collection, thinking the vapors might neutralize the acid. (I'll let you guess how putting a hypergolic liquid that reacts with air around flammable paper turned out.) Diethylzinc has a very low flash point, only a few degrees above room temperature, and remember: it's the vapor that catches on fire, not the liquid. With that much vapor, it's an explosion waiting to happen. Its flames are blue rather than green, however, and it has the same issues as trimethyl borate. So, although it is possible that wildfire has hypergolic properties, there are currently no known hypergolic liquids that exhibit all the characteristics of wildfire.

GREEK FIRE

By now you are probably screaming at me, asking why I haven't yet talked about Greek fire, the obvious precursor to wildfire. I held off because I wanted to first talk about what chemists actually know, not about Greek legends. But, as expected, I got to Greek fire eventually. For those of you who weren't cursing me in their heads while waiting for me to get to this part, here is a brief history of what Greek fire is and why it is so important to this discussion. Greek fire wasn't used by the Greeks but by the Byzantine Empire, which is also called the Eastern Roman Empire. It managed to outlast the Western Roman Empire by about 1,000 years with the help of Greek fire. The Byzantines never called it "Greek fire" but rather sea fire, liquid fire, or median fire. The name can possibly be attributed to its inventor, Kallinikos, a Syrian refugee referred to as a Greek by early Byzantine accounts. Greek fire had the same properties of fictional wildfire with the exception of the color; there are no reports about Greek fire being green. The key features were its ability to burn on water, the fact that it could only be extinguished by vinegar or sand, and its unique deployment method, the flame torch. It's extremely difficult to separate fact from fiction when it comes to Greek fire, but I'll do my best. The methods for making and deploying Greek fire were a state secret and have been lost to the ages, but, lucky for us, there are fragments of texts that remain and can shed light on its use as a weapon. Unlucky for us, these sources don't really agree with each other from a scientific perspective.

There are a handful of historians and chemists who have researched the composition and deployment of Greek fire by examining surviving texts and artifacts such as stone pots. What no one disagrees on is the fact that Greek fire single-handedly turned the tide of the Arab–Byzantine wars. It made its first appearance sometime around 672 AD in a similarly desperate situation as the battle of Blackwater Bay, only with a slightly longer time line. The city of Constantinople had been besieged for two years by the Muslim Arab fleet. Arab forces had conquered a sizable chunk of the Byzantine Empire at this point, and they were heading for the capital. Luckily for Constantinople, they were given their own Tyrion in the form of Kallinikos. All sources agree he was responsible for giving "liquid fire" to the Byzantines. Some have suggested that he didn't invent it but simply brought it over from Syria. Seeing as the Arab army was defeated using the technology, I'm going to guess they didn't actually have some of their own. Kallinikos created a deadly fuel and a deployment system, called a *siphon*, that turned it into a fleet-destroyer. Armed with this new weapon, Constantine IV broke the two-year siege and started a legend. GRRM couldn't have written it better himself; in fact, I think history wrote that scene specifically for him.

What was the magic that changed the Byzantine fleet into the most feared navy of the time? What had this alchemist Kallinikos created? The definitive work on the subject is *A History of Greek Fire and Gunpowder* by J. R. Partington.[6] This fabulous book traces the history and chemistry of incendiary weapons throughout the ages. Partington was a chemist and a historian, and this book approaches its subject through the lens of both disciplines. Bert S. Hall, who wrote the introduction, thought the book didn't tell a good enough story and focused too much on the chemistry. I, as a scientist, didn't care as much for the story and wanted more of the chemistry, so he must have struck a good balance. Alex Roland wrote an article focused on the technology and secrecy involved in the making of Greek fire and made an interesting point in noting the difference between engineering and technology that's worth highlighting and still seems to hold true.[7] *Engineering* is used to solve a problem, and advances in engineering are published and shared with the world. *Technology*, by contrast, is inherently a secret endeavor, and the results of any technological advances should be hidden at all costs so that they benefit only those making the advances. Engineering is published in journal articles; technology gets patents. Greek fire was very much war technology. It is helpful to think of Greek fire not just as a flammable compound but as an entire weapons system. In the Battle of Blackwater, the delivery system was pretty straightforward: dump the fuel on the water and light it with a flaming arrow. Greek fire was not that simple and involved the construction of a system to light and launch the fuel using a flamethrower, and because the attacks were being launched from ships, the system also had to be stable enough that it didn't light the ship on fire by accident. The theory is that to keep the secret of Greek fire, no one person or group of people knew all the steps it took to create it. Those people who could build a siphon couldn't operate it, whereas those who could create the actual fuel had no idea how it would be deployed, and so on. The composition and deployment of Greek fire changed over the years from its debut in the siege of Constantinople until the fall of the Byzantine Empire. There is much debate about its composition, with one main sticking point being the inclusion of saltpeter.

Partington, with all his fair-mindedness and his reviews of many sources, is adamant that there was no saltpeter—the main ingredient in gunpowder—in Greek fire. He's alone in that belief, however; practically everyone else thinks there must have been. Eyewitness accounts say that Greek fire made a sound like thunder and trailed smoke, both hallmarks of gunpowder; however, gunpowder is commonly believed to have been invented in the ninth century in China. Saltpeter, or potassium nitrate, is easily found in nature was the key to modern explosives. The molecule has three oxygen atoms that it is happy to give away. For a fire to burn, it needs oxygen, but a mass of fuel usually only gets oxygen at the surface—not enough for a satisfying explosion. Mixing in saltpeter as an oxidizer gives all of the fuel access to oxygen and causes it to ignite all at once when heated. This was the key to creating gunpowder and, many argue, Greek fire's explosive power, smoke, and noise. With the exception of Partington, historians of chemistry agree that there must have been some sort of oxidizer in Greek fire. Given saltpeter's abundance and effectiveness, it is the natural suspect. The main argument against it, from Partington's perspective, is that there is no record of it in any other civilization until 1000 AD in Asia, and that it wasn't exactly needed to create effective Greek fire. Literature reviews would indicate that researchers have agreed to disagree—as in, everyone agrees except Partington. Saltpeter controversy aside, there are several ingredients that are universally accepted as likely components of Greek fire: petrol, pine tar or resin, and sulfur. Some argue that quicklime, which catches fire in water, was also included, but Partington argues that the properties of quicklime were nothing new and that it was not needed to produce the flames. One historian, Anna Komnena, wrote about the composition of Greek fire around 1100 AD, and there was no mention of quicklime. She said the material was produced with petrol, pine resin, and sulfur, and Partington concludes that those three would be enough to create the deadly fire. The petrol acted as a fuel and burned on water, the pine resin acted as a thickening agent, and the sulfur added the ability to ignite in water. Kallinikos had created something very, very similar to napalm about 1,300 years before Dow Chemical got around to it. Some accounts say that naphtha, a distilled version of petrol, was used instead of straight petrol. Indeed, the one big difference seems to be that napalm is difficult to light and by all accounts Greek fire could be lit with normal torches.

The fuel composition wasn't the only important piece of this wartime technology. There needed to be a way to deploy the mixture effectively. This is where GRRM's account diverges significantly from historical ones. The Byzantines created a device called a siphon that heated and pressurized the fuel compound before blowing it out of tubes and onto enemy ships. The tip of the tube would have a small flame that would ignite the fuel as it passed over it. The modern-day equivalent would be putting napalm in an Instant Pot with a valve, tube, and nozzle output, and a candle at the end. The contents were pressurized using heat and a bellows. The heat not only increased the pressure but helped the resin dissolve into the petrol. Once the mixture was pressurized, the valve would be opened, and the pressure would force the fuel out of the end, creating a flame torch. The nozzles would be on swivels so that the flaming material could be aimed in any direction. I am a member of the cult of the Instant Pot, and even though I know it is irrational, I always feel a little nervous about having a highly pressurized rice bomb in my kitchen. I can't even imagine the danger of pressurized napalm in a primitive bronze vessel on a wooden ship. The Byzantines, however, seemed to have more success than failure: there are no records of a ship blowing up due to the siphon system backfiring. It should be noted, though, that people usually don't record their failures. One can assume this wasn't a system with a 100% success rate, but it was effective enough to change the face of naval warfare.

Whether or not wildfire is indeed the same as Greek fire, it's very easy to conclude that GRRM certainly intended it to be so. However, there's the color that is still an issue. Of the likely components—petrol, sulfur, pine resin, saltpeter, and maybe quicklime—none burn green. Sulfur and saltpeter, however, both burn with blue flames, and petrol and naphtha burn yellowish-orange, so Greek fire most likely burned with a bluish tinge. There are two ways to look at this: either wildfire had some additional component to it, most likely a boron compound, or GRRM should have described wildfire as being a vivid blue. From a narrative perspective, I can see why he went for green flames over blue—we've all seen blue flames on a Bunsen burner or particularly hot candle, but rarely have we seen green flame. That green color certainly adds to the dramatic effect; however, it keeps getting in the way of solid scientific explanations.

Incendiary weapons have turned the tides of war on several occasions. Their use prevented the fall of Constantinople to the Arab forces and helped the Byzantine Empire survive another 1,000 years after the fall of Rome. Backlash from the devastation they caused led the US military's withdrawal from Vietnam. And in GRRM's Westeros, they kept Joffrey on the throne, for better or worse. It is more than plausible to have a fire capable of the kind of destruction wildfire is known for. There have been several instances in history where such materials were made. Greek fire, though lost to history, is by far the closest. The key difference between it and napalm B is that it is very easy to light. The big downside is that to use it effectively you have to essentially put a pressurized, napalm-filled bomb on your own ship. The strategy employed by Tyrion in the Battle of Blackwater is more than plausible, however, and Bronn could have easily ignited the fuel with an arrow. Napalm B seems much more closely aligned with the mythical properties of wildfire described in the book, but lacks the easy lighting of Greek fire. Both of these are compounds that the Westerosi alchemists could have created, though napalm B is a bit of a stretch. It might be possible that boron or another element was mixed in with either Greek fire or napalm specifically to create the terrifying green color. It's not unheard of to do things like that simply to intimidate enemies. What I find interesting is that the real winner of the Battle of Blackwater wasn't wildfire but Tyrion's strategy. He could have pulled off the same tactic with a petroleum product, an empty ship, and Bronn's arrow. The wildfire was really secondary. It was a spectacular show, but as the Cuyahoga River showed us, lots of things can burn on water on a large scale. Though legend will always say it was wildfire that won that battle, it was really Tyrion with his drinking and knowing things.

As a quick side note, I have both Global Entry and TSA PreCheck memberships, and I had never been selected for the super special security screening at the airport until I started writing this chapter. Don't search for "how to make napalm" unless you want to get to second base with the TSA.

11

Houses Targaryen and Lannister

The Genetics of a Family Tree with Few Branches

The seed is strong.

—Jon Arryn, *A Game of Thrones*

One of the constant jokes about *Game of Thrones* fans is how their reaction to incest in season 1 is "Ewwwwww," but by season 7, it's evolved to "Woohoo, they finally got together!" Although fans' attitude toward incest may have changed, the consequences of incest remain the same. In Westeros there seems to be quite a few of them. Jaime and Cersei Lannister had three children, two of whom turned out pretty OK and one of whom, well . . . not so much. Was Joffrey's madness inherited from his questionably sane mother, or was it a side effect of his dad also being his uncle? Considering Joffrey's dad being his uncle was a pretty big deal to the current inhabitants of King's Landing, we can assume this was not something normally done. The Targaryens, by contrast, had no problem whatsoever with marrying their siblings. It's strongly hinted that the Targaryens' inbreeding was the cause of the Mad King's insanity. But does science agree? Could generations of inbreeding lead to crazy offspring? Inbreeding in royals is nothing new or even fictional; many of us learned about the Habsburgs and the pharaohs in history class. There are real-life royal examples of the problems inbreeding can cause, from hemophilia to jaw deformities. Incest taboos are so strong today that many people have a hard time even backing away enough to look at the science of why having kids with your sister is a bad idea. Attraction clearly plays a role in the *Game of Thrones* incest. Many (most?) people can't understand how a brother and sister could be attracted to each other. But that raises the question of why, exactly, someone does not find their sibling attractive. As a scientist, I found the answers pretty interesting. I hope you can put aside your repulsion to the notion of incest long enough to enjoy learning why you feel that way in the first place.

How Genes Work

As you can see from my photo on the book's jacket, I have blond(ish) hair and blue eyes. You might guess, then, that my mother or father or both have blond hair and blue eyes. That's how genetics works—you look like your parents. Your guess, however, would be wrong. My dad has blue eyes and black hair. My mom's eyes were hazel, and I'm not sure even she remembered what her original hair color was, but it certainly wasn't blond. Genetics is more complicated than just looking like your parents. You can inherit some traits but not others; some are dominant, some are recessive, and some are linked to sex. In 1856, a farmer-turned-friar named Gregor Mendel decided he wanted to find out a bit more about how inheritance works. This is the simple version that hopefully you have a hazy memory of from high school, but I'm going to go through it again since this model is a pretty good way to eventually look at what might have happened to Joffrey.

Initially, Mendel conducted his experiments with rats, but the bishop found such experiments unseemly for an abbey. Mendel turned to plants, organisms that propagated through sexual reproduction, but slightly less graphic sexual reproduction. (Not a lot of plant porn out there.) At that time, it was possible to buy what were called "true-breeding" seeds that would reliably produce pea plants with certain characteristics. For example, if you bought true-breeding seeds for tall pea plants, the resulting pea plants would be tall no matter how many generations you bred. Mendel selected seven different traits for which he could get true-breeding seeds: seed shape, flower color, seed coat tint, pod shape, unripe pod color, flower location, and plant height. He wanted to see what would happen if he crossbred the plants in a scientifically rigorous way. The prevailing idea was that crossbreeding tall pea plants with short pea plants would result in pea plants of a height between the two, but that's not at all what Mendel found. What he found was that if you crossbred tall plants with short ones, the resulting plants were tall. All of them. He theorized that instead of these traits mixing, the tall trait was "dominant." He then wanted to see what would happen if he bred the resulting tall pea plants with each other. Without knowing what these plants' parents looked like, one would assume the baby plants would all be tall. That wasn't what he found, however—of the second generation of plants, three-fourths were tall, as expected, and one-fourth were short. When he bred all the short plants with other short plants, the resulting plants were short, but breeding second-generation tall plants with each other also led, again, to a number of short plants. This was the case for each of the seven traits he observed. One of the characteristics was dominant, and one was hidden but passed along to the next generation. The trait—height, color, seed shape, and so on—was eventually called a *gene*, and the characteristic, or type, of that gene—tall, short, purple flower, white flower, and so on—was eventually called an *allele*. Mendel theorized that gametes, the cells responsible for sexual reproduction, each have one allele. A sperm cell, for example, will carry either a tall allele or a short allele, but not both. When the gametes come together, the resulting plant will have two alleles, one from each parent plant. What the resulting plant looked like was called the *phenotype*, and the type of alleles it had was called its *genotype*. A plant with two different alleles, such as a tall pea plant that was a carrier of the short allele, was called *heterozygous*, and a plant with the same alleles was called *homozygous*. Mendel's work was not recognized for what it was until the early 20th century. The terms "gene" and "allele" weren't even coined until after Mendel's time but are now standard in discussions of Mendelian inheritance. There was some controversy around who "rediscovered" (and in one case did not credit) Mendel's work, but that rediscovery was the birth of modern genetics.

Laws always seem to come in threes in science, and there's no exception here. The first is the *law of segregation*, which says that when the gametes are formed, the alleles for each gene are split, with only half going into each gamete. Each gamete has only one allele for each gene. A plant gamete can't have, say, both short and tall alleles but no alleles for flower color. The second law, the *law of independent assortment*, says that genes are independent of each other. The genes for height aren't linked to the genes for flower color or seed shape. This one gets a little murky because there are genes that tend to be linked and inherited together. The traits Mendel picked, however, were not linked, so the law of independent assortment worked in that case. The third law is the *law of dominance*, which states that one allele is dominant and if that allele is present it will be expressed in the phenotype. Much as Newton's three laws of motion explain most of the physics you see in everyday life, Mendel's laws of inheritance describe the fundamentals of genetics, enough to give you a pretty good handle on how things generally work. And like Newton's laws, they have since been built upon and contradicted, with scientists finding exceptions to the rules upon further research. The goal of this chapter, however, is to understand what happens if you procreate with a very close relative, so a working knowledge of Mendelian inheritance will serve just fine.

When using Mendelian inheritance principles to figure out the statistical probability of an offspring's genotype and phenotype, doing it in your head can be a little complicated. Enter Reginald Punnett. Punnett worked with the geneticists who rediscovered and built on Mendel's work. He realized that the easiest way to look at the complicated percentages involved with Mendel's laws was a visual representation that came to be known as a Punnett square. I'm sure many of you remember solving these puzzles in biology class. In a Punnett square, the father's genotype goes in the boxes on the top and the mother's genotype goes in the boxes on the side. The boxes in the middle are then filled in to show what would happen if the gametes had those specific alleles. Describing it in words is hard, which is exactly what Punnett found out and why his squares exist. Table 11.1 shows a Punnett square for the flower colors of a pea plant if both parents are heterozygous.

TABLE 11.1

	B	**b**
B	BB (purple)	Bb (purple)
b	Bb (purple)	bb (White)

You can see that there is a 75% chance the offspring will inherit the dominant allele for purple flowers and a 25% chance they will only inherit the allele for white flowers. These are precisely the statistics Mendel saw. According to Mendel's laws, genes are independent, so it's possible to do even bigger Punnett squares for more than one gene at a time. A Punnett square displays not only the phenotype but also the genotype. If you want to see what happens in subsequent generations, you can take the genotypes from one square and continue to the next generation. All of Mendel's laws can be summed up in Punnett's neat boxes.

How Traits Are Passed On

Mendel came up with the statistics that determine which traits are inherited, but he did not tell us *how* those traits were inherited. The terms "dominant" and "recessive" became common parlance, and statistics could be used to predict the characteristics an offspring was likely to have, at least for some rudimentary traits; however, no one had yet figured out the mechanism by which the traits were passed down. By the mid-20th century, many scientists were trying to understand the role of DNA in sexual reproduction and heredity. In 1953, James Watson and Francis Crick made an amazing discovery: Rosalind Franklin's data. (I know this joke is an internet meme, but I had to include it.) Or at least that's the story that we've all heard. In reality, the search for a mechanism by which traits were passed along is a bit more complicated than we've been led to believe. Yes, Franklin, Crick, and Watson took the data and proposed the model for the structure of DNA, thus giving insight into its function, but the more important question is: Why did anyone care? We figure out molecular structures all the time. What was so special about this molecule?

DNA, or deoxyribonucleic acid, was first isolated by Swiss physician Friedrich Miescher in 1869 while he was trying to classify proteins in pus from bandages. He found something that was not a protein but rather an acid housed inside the nucleus of white blood cells, and called it a "nuclein." Over the course of 20 years of research (1909–1929), Phoebus Levene, the head of the biochemical laboratory at the Rockefeller Institute of Medical Research in New York City, figured out what the molecule was made of, though he assumed it was much, much smaller than it was. William Astbury was actually the first to take pictures of DNA, though they only showed that the molecule had a regular pattern, not what that pattern was. By 1937, scientists knew that the cell nucleus contained an acidic molecule, what that molecule was made of, and that it had a particular regular shape, but none of that hinted at what the molecule actually *did*.

Two more steps were crucial in understanding the function of DNA: theorizing a way for genetic material to be passed along, and then showing that the nucleic acid was responsible for that task. Nikolai Koltsov, who is rarely mentioned in connection with DNA, was the first to propose that traits were most likely passed along by a "giant hereditary molecule" made up of "two mirror strands that would replicate in semi-conservative fashion using each strand as a template." Step one complete—a model for inheritance theorized.[1] Three experiments were crucial in understanding that the nuclein was that molecule. The first was carried out by Frederick Griffith in 1928. He showed that if you mixed live, nonlethal bacteria with lethal bacteria that had been killed, rats would still get sick. There must be some mechanism of "transforming" the live bacteria into deadly bacteria. Building on this, Oswald Avery, Colin MacLeod, and Maclyn McCarty proved in 1944 that DNA was responsible for that transformation. Erwin Chargaff filled in the gaps by showing that different species have different DNA. No matter what species, however, DNA always contained adenine (A) and thymine (T) in the same amounts and guanine (G) and cytosine (C) in the same amounts. As you may remember from high school biology class, these are the nucleotide *base pairs* that fit together to make the full strand of DNA. By this point, all the pieces were there—the key was putting them together.

What Franklin, Watson, Crick, and their colleague Maurice Wilkins ultimately did was unlock the keys to the structure and mechanism of DNA as the genetic code. *Photo 51* showed DNA's infamous double helix (or twisted ladder) structure. I could spend an entire book talking about the drama of the discovery, and indeed many have, so I'll cut to the punch line. Their model showed on a molecular level how traits can be passed from one cell to another and from parents to children. DNA is a double-stranded molecule with bases that pair only with a specific other base, and the order of these base pairs is different in every single organism. In 1958, Crick published what is now called the central dogma of molecular biology: DNA can replicate and be passed from one cell to subsequent cells before turning into RNA, which creates proteins that create organisms. (I'll spare you the replication details, as I'm sure you saw the video in 9th grade.) Your body, everything from your hair to your blood, is governed by proteins and protein interactions. Curly hair has different proteins than straight hair and some proteins lead to attached earlobes and some don't. At least that's the simple version. Of course, in reality, it is much more complicated.

23 AND YOU

DNA is grouped into *chromosomes*. Each chromosome is a full strand of DNA, and if everything goes according to plan, humans have 23 pairs of chromosomes. These chromosomes make you *you*. They say whether you are going to be tall, short, brown-eyed, blue-eyed, a hemophiliac, or anti-cilantro. What makes it much more complicated is that genes can be turned on and off, environmental factors can change how a gene is expressed, chromosomes can be added or missing altogether, and sometimes genes combine in weird ways. It's not as easy as intro to biology would lead you to believe. But intro to biology is a pretty good place to start.

A chromosome is a single molecular package. It's a strand of DNA wound around support proteins called histones, arranged in the hallmark "X" shape. The center part of the "X" is called the centromere. Most cells have 46 chromosomes arranged in 23 pairs, but reproductive cells—eggs and sperm—have only 23 chromosomes, just one half of the pairs seen in other cells. When they come together and the egg is fertilized, the embryo gets its 23 pairs. Well, if it's a girl it has a full 23; if it's a boy it has 22.5. Boys have the Y chromosome, which is a fancy way of saying half an X. It's the missing half that leads to what are called *sex-linked traits*. In sex-linked traits, male offspring don't have two copies of some genes—specifically, the genes that are on the missing half of that last chromosome. Because there aren't two options, whatever gene is on the bottom half of the X chromosome in the pair is the gene that's expressed. Women can be carriers of a mutation because if there's a problem they have a backup option. Color blindness is a great example of this. For a woman to be color-blind, she'd have to have two messed-up copies of the gene, whereas men only need one messed-up gene to be color-blind.

How do these chromosomes and the proteins they produce make you look like your parents or have hemophilia? Until now, I've talked about genotypes and phenotypes as if there's a one-to-one relation between the two; that is, if you have the genes for blue eyes, you have blue eyes. In reality, it is far from being that simple. Sex-linked traits are just the beginning. The proteins created from your DNA don't operate independently—they work with each other. Your phenotype is determined by how the proteins in your genotype interact. Figuring out exactly how a particular genotype produces a particular phenotype is extremely complicated, and it is currently an active area of research. In particular, scientists are studying how various genotypes can present as diseases, and how environmental factors influence how and which genes are expressed (part of a field called *epigenetics*).

Identifying Genetic Diseases with the Help of Incest

Finding the gene responsible for a particular trait is not that easy. Scientists are smart and all, but they can't look at someone's DNA and say, "Aha! I see the problem! It's right here!" To find the gene responsible for a specific disease, a good place to start is by examining the DNA of a person who is known to have that disease and figuring out which gene encodes a protein that causes a particular trait. If you take into account how much DNA the average person has, this is like trying to find a needle in a field of haystacks. Instead, scientists approach this challenge by looking at groups of people who have a particular disease of interest and comparing their DNA. This is called *positional cloning*. The DNA of people who have the disease is compared with the DNA of people who don't. By looking at the differences between the two populations, it's possible to figure out which genes code for the disease. The first gene to be identified this way was the gene for cystic fibrosis.

When trying to link a gene to a disease, scientists look at families where the disease has been unambiguously diagnosed in a subset of the family members. From there, it's a matter of statistics. In the case of diseases encoded by a single gene, it's pretty clear to see if someone has the disease, and if the gene is passed through simple inheritance it is quite easy to track down which section of DNA is responsible. This is how the gene for cystic fibrosis, pinpointed in 1989, was identified only 26 years after the diagnostic criteria were first laid out. The gene that codes for Huntington's disease, identified in 1993, was also found quickly with relatively few patient samples needed. The genetic causes of these diseases took less time to identify than others because they are both caused by one mutation to a single gene. The cystic fibrosis gene is recessive and the gene for Huntington's is dominant, but both are passed through simple inheritance. A basic Punnett square can determine the probability of inheriting a particular genotype and phenotype. When the same outcome is consistently achieved, conducting additional statistical assessments doesn't add any new information.

However, if a disease is caused by mutations in several different genes, has a complicated inheritance structure, or is frequently misdiagnosed, it is much more difficult to pinpoint the cause, and a correspondingly larger group of patients is required to statistically identify the cause of the disease. If finding the mutation is not so straightforward, there are still statistical ways to find a specific gene. In the case of BRCA1, which can have mutations that predispose someone to breast cancer, it was difficult because very few breast cancer cases (5–10%) are attributed to BRCA mutations. If you had the mutation, however, you were far more likely to get breast cancer, and often at a younger age. To separate the signal from the noise, researchers studied the DNA of people who developed breast cancer at a younger age. By looking specifically at that group, the researchers were able to pinpoint a mutation that was shared by a large number of patients—enough to safely conclude that the mutation could cause breast cancer.

Because statistics plays such a large part in finding the specific gene responsible for a disease, positional cloning can be particularly difficult for rare diseases. The Lannisters could help with that. Because the likelihood of having a rare disease drastically increases in cases of incest, looking at the DNA of those afflicted with a disease that also happen to be extra related can speed up locating the gene. This is called *homozygosity mapping*. In this method, the DNA of two related individuals who have the disease is compared to a common relative with the disease. It can often be difficult to pinpoint a mutated region responsible for a disease because each mutation is different; however, since the patients with the mutation inherited it from the same person, both of them should also have that mutation, which makes it easier to spot. Joffrey's DNA could have illuminated some scientific mysteries. This method is very useful when there is more than one mutation responsible for a disease; since 1995, there have been about 200 published studies in which consanguineous individuals helped find the mutations that cause rare diseases. Incest for the science win![2]

THE IMPORTANCE OF VARIETY AND THE PROBLEMS CAUSED BY THE LACK OF IT

After that more complicated discussion of how a genotype becomes a phenotype, let's go back to the relatively simple Mendelian genetics for the next discussion. This is a simplistic overview of how a lack of genetic diversity produces problems, but it's a good first pass and sufficient for providing a scientific basis for how messed-up Jaime and Cersei are. From an evolutionary perspective, the goal of any organism is to grow up and make more of itself. I know this isn't always true for humans in specific cases, but on the whole, people want to grow up and make more humans—it's one of our most basic and undeniable biological instincts. To do that successfully, an organism's phenotype has to allow for it. A person needs to be attractive enough, able to provide for the kid, relatively free from disease, and fertile, just to name a few necessary qualities. But things don't always work out that way. Let's say that in a population there is a gene with a recessive allele that leads to sterility. Probably only a small percentage of people would have it—maybe 1 in 1,000. In almost all cases, that person has a dominant allele that takes precedence, so being a carrier of the recessive allele doesn't matter much. If two random people were to meet and have kids, there's only a 1 in 1,000,000 chance they're both carriers of the infertility allele. And even if this one-in-a-million chance occurs, for the most part the kids will be fine since there's a 75% chance they'll get at least one copy of the dominant gene. The odds are really good that one parent will be homozygous and have two good dominant alleles to pass along. Even if the other parent is a carrier, all the kids will be fertile and only two out of four will be carriers. If all four of those kids go out into the wide, wide world and find partners, there is a really good chance that the partners won't be carriers. Statistically, this means there's a really good chance that the grandchildren will also be fertile. If there's a huge gene pool in which to fish, the chances are good that the resulting phenotypes will have the right stuff to continue reproducing.

Now for another example that doesn't turn out as well. Let's assume there's a recessive allele for a birth defect, maybe a heart problem, and there is a parent who is a carrier. Let's name that parent Tywin. Tywin goes out and has kids with someone who isn't a carrier; let's call her Joanna. Table 11.2 shows the resulting Punnett square. Tywin is across the top with a dominant healthy heart gene and a recessive heart problem mutation. Joanna, on the left, has two copies of the healthy gene.

TABLE 11.2

	H	**h**
H	HH	Hh
H	HH	Hh

You can see that there is a 50% chance of their kids being carriers. If those kids go out and find partners who are not carriers, and the odds are really good that a random partner won't be a carrier, all is well. The grandchildren might be carriers, but they won't show the defect. Now, let's assume the kids are named Cersei and Jaime. They don't particularly need to look much farther than the room down the hall. Since their dad was a carrier, there's a 50% chance either of them is a carrier and a 25% chance they both are. There's now a 6.25% chance that their kid will have a heart defect. Those are some pretty bad odds. By keeping it in the family, they have drastically increased the odds that their child will have a birth defect. To make it worse, the heart defect is just one of many, many possible recessive defects. By refusing to date other people, they have increased the probability of having children who could exhibit a huge array of problems.

The Targaryens were also very fond of their siblings, but they were open and proud of their choices. In their eyes, by not usually marrying outside the family they were not introducing weaker genes into the lineage. In scientific terms, they believed there were none of the recessive genes that cause problems in their family's DNA. In some cases, they were right about not introducing outside influences. Blond hair and light eyes are recessive traits. If House Targaryen were to marry outside the family, and to marry Starks or Baratheons in particular, it's very likely that their children would not retain the trademark Targaryen coloring. This is what Jon Arryn meant by his dying words, "The seed is strong." He was saying that brown-haired Robert, who comes from a long line of brown-haired people (in Mendel's terms, he was a true-breeding brunet) and who clearly has no hidden recessive blonde gene, was not going to produce three blond children. This is also why Jon Snow (or Aegon Targaryen, depending on what season you're watching) has dark coloring even though his father was quite blond. Starks being true-breeding brunets, just like Baratheons, there were no potential recessive blonde genes for Lyanna to pass along to Jon. Jon had to have dark hair. It's also worth noting that the Targaryens were marrying siblings for many generations and only toward the end with the Mad King did issues seem to pop up, or at least that's what we're led to believe. They may very well have had few detrimental recessive genes in their family genetics. However strong a family may think their line is, statistically, the Targaryen way is not usually a chance worth taking.

While talking about this chapter with some friends, I had someone ask if inbreeding between identical twins would be worse than inbreeding between a normal brother and sister. You may have seen the problem with this before even finishing that sentence, and I'll leave it as an exercise for the reader to point out the absurdity of the question.

Finding the Balance between Alike and Different

All of this discussion of the biological perils of inbreeding seem to imply that the best thing to do is to find someone as different from yourself as possible. It turns out this is not what humans tend to do. It's also not necessarily the best idea. As much as I've talked about the importance of genetic diversity in successfully passing on one's genes, it is equally important to find genes that complement one another. Much of this chapter has discussed the issue of *inbreeding depression*, meaning that offspring are less fit than their parents. In other words, inbreeding is a bad idea and the kids will be messed up; however, it is not fair to say that inbreeding depression is the only possible issue. There is also a phenomenon called *outbreeding depression,* which means that when parents have greatly differing genes their offspring are also less fit. Sewall Wright argued in 1933 that there was a balance between inbreeding and outbreeding.[3] As with extremes of inbreeding, it's conceptually easiest to start thinking of the extremes of outbreeding. One extreme of outbreeding depression is the mule, which is a cross between a horse and a donkey. Although it may be a great work animal, it is sterile and therefore not very "fit" in terms of genetic survival. There are passing mentions of potential outbreeding depression in animals in other sources, but there are few easily found studies that specifically address the fitness of offspring from extremely genetically diverse parents. In Patrick Bateson's book *Behaviour, Development and Evolution,* however, he points out that there is a good conceptual case against excessive outbreeding. He gives the example of a jaw and teeth. It would be a bad idea for someone with a small jaw and small teeth to have kids with someone who had a large jaw and large teeth. If jaw size and teeth aren't genetically linked it, could end in offspring with teeth that don't fit their jaw and therefore loads of eating problems.

If we can't marry our close relatives and we can't marry people who are nothing like us, what's the optimal balance? Turns out, believe it or not, it's your third or fourth cousin. A group from Iceland looked at the fertility of Icelandic couples from 1800 to 1965. Iceland is particularly well suited for these types of studies since it has an extremely homogeneous population. There is little variation in culture, socioeconomic status, and birth control use, and both the immigration and emigration rates are low. The group looked at how many kids a couple had and then how many grandkids and great-grandkids they had. It was normalized by the average number of children per couple. They then examined how closely related each couple was. The researchers used the database of an Icelandic company called deCODE, which collects genetic information from the entire population of Iceland with the goal of linking genotype to phenotype. Luckily for them, the population of Iceland was only about 270,000 when they started. The database is extensive and allowed researchers to accurately determine how closely related couples were. What they found was that couples who were third or fourth cousins had the most kids, grandkids, and great-grandkids. If couples were more or less related than that, their fertility was lower. This trend held for every 25-year interval over the 165 years studied. No matter the population movement, average family size, economic conditions, and other factors that changed over 165 years in Iceland, people who married their third or fourth cousin had more kids, and more kids who were able to have kids, than people who married closer or more distantly related relatives. Looks like they struck a good balance—close, but not too close.[4]

How does attraction play into all of this? Although there can be many cultural reasons to get married and have kids with someone you are not attracted to, from an evolutionary perspective we should be naturally attracted to people who we'd be most likely to have healthy kids with. So, are we attracted to people who look like us, or at least look like they might be a third cousin? In the beginning of the 20th century there was a disagreement between a Finnish sociologist named Edvard Westermarck and everyone's favorite psychologist, Sigmund Freud, because it's impossible to talk about incest without including Freud. Freud argued that incest taboos were in place because we are naturally attracted to our close relatives and without the social taboos in place there would be detrimental inbreeding. Westermarck hypothesized that if you grew up with someone from birth to roughly age six, you wouldn't be attracted to them. It was a kind of reverse sexual imprinting that would later be known as the Westermarck effect. In 1994, A. P. Wolf set out to test Westermarck's hypothesis by studying marriage statistics in Taiwan. In one marriage tradition, the intended bride was adopted at a young age and grew up with her future groom. In another marriage tradition, the partners met later in life. Wolf found that the first type of marriage produced fewer children and had higher rates of both divorce and marital infidelity. Similarly, in Israeli kibbutzim, where children, some related and some not, grew up together, few went on to marry each other later in life. It appears that being raised together might be why we don't find our siblings attractive.

Modern research seems to imply a balance between the two ideas. We are attracted to people who look like us, but not to people who were raised with us, and if we realize we're related to someone we will automatically rank their attractiveness lower. This is, of course, an oversimplification, since human attraction is pretty complicated. Regardless, it can at least be explored a little bit. Chris Fraley from the University of Illinois and Michael Marks from New Mexico State University published a three-part study in 2010 that explored how attracted people might be to their family members and to people who resembled them.[5] In the first part of the study, participants were shown a picture of their opposite sex parent quickly—so quickly they didn't even know it had been shown to them—and then shown a picture of someone of the opposite sex. They were asked to rank this stranger's attractiveness. The group that was covertly shown a picture of their parent ranked the subsequent images as more attractive than the group that was not shown their own parent. This seemed to support the Freudian model of attraction. The second experiment tested whether or not people were attracted to people who looked like them. Participants were shown 50 faces and asked to rank the sexual attractiveness of each face. What they didn't know was that some of the faces had been morphed with the participant's face to varying degrees from 0%–45%. A control group was shown the same faces, but they weren't looking at faces mashed with their own. Instead, they were looking at faces mashed with other participants. Researchers found that people rated the faces morphed with their own as more attractive than faces that weren't. Morphed faces seemed to be more attractive to everyone, but were much more attractive to those who were looking at their own face. This seems to indicate that people prefer people who look like themselves, which also supports Freud's theory. In the last experiment, people were told that some of the faces they would see would be their own face morphed with a stranger. Enter the incest taboo. People consistently ranked the morphed faces lower than those in a control group who saw the same faces but were not part of the images. People did not want to rank themselves highly. These three experiments support Freud's hypothesis, but they're also based on a slight misinterpretation of Westermarck's original work. Westermarck demonstrated that being raised with someone means you are much less likely to find them attractive. The research was done in communal living situations where the children weren't related and probably didn't look like each other. Still, on the whole, they were not attracted to each other. There's enough room in these theories for both Westermarck and Freud to be correct. We are attracted to people who look like us but not if we grew up with them. It would seem, however, that Cersei and Jaime proved them both wrong.[6]

On the flip side, Theon Greyjoy does a great job supporting both Westermarck's and Freud's theories. When Theon goes back to the Iron Islands, he immediately (and unknowingly) tries to hook up with his sister Yara (or Asha, if you're a book person), and both the audience and Theon are grossed out when they realize the two are related. This is a great example of all the aspects of incest taboos and attraction coming into play. He wasn't raised with Asha, so he was open to being attracted to her; she looked like him, so he found her attractive; but then he realized who she was and was more than a little nauseated due to the cultural incest taboos referenced by Freud. Unknowingly hitting on his sister is probably one of the less repulsive things about Theon, however.

As a physicist, my takeaway from all of this research into attraction is that psychology and human interactions are hard, and I'm glad I chose to study physics, which involves neither. Also, I would strongly caution against attending a family reunion with relatives you've never met before. Obviously, taboos would prevent anything from happening, but there is a strong possibility you'd end up leaving with a lot to discuss with your therapist.

DIAGNOSING A MAD KING

In *Game of Thrones*, there is some serious incest: brother and sister (and twins, at that), father and daughter, and now aunt and nephew. The gene pool for many of the noble families is not Olympic-sized but rather like a backyard hot tub. To some degree, and with the Targaryens in particular, it could be argued that the detrimental recessive genes that are usually present in repeated inbreeding may have died out in the line; that is, those carrying the genes for infertility or diseases could not reproduce or died before reproducing. But this applies only to genes that would cause an untimely death, not genes that would cause madness. Could madness run in House Targaryen? And what about Joffrey Baratheon (who's really a Lannister)? How did he end up not only vicious but irrationally so?

Controversial though it may be, research suggests there may be such a thing as a "psycho gene": the MAOA gene. This gene encodes proteins that break down serotonin, dopamine, and norepinephrine. Basically, the gene makes a protein that eats up the brain's feel-good neurotransmitters. A mutation of the MAOA gene leads to these neurotransmitters not breaking down, and this causes aggression.[7] The gene is a sex-linked trait. This means that men only have one copy of the gene and women have two copies. If there are two copies and at least one works, then the phenotype of aggression is probably not seen. To make it a bit more complicated, the gene expression is affected by how a person is brought up. If they are raised in a stressful or abusive environment, they are more likely to exhibit aggressive behavior. This is controversial because the expression of the gene is by no means an inevitability, nor is it an excuse for aggression. It is more likely to show itself when the person is raised in a turbulent household, but again, it is not inevitable. It's worth looking at how this might play out in House Lannister, but I need to caution you that this is a very rudimentary analysis and should only be taken as the genetic equivalent of "Assume a spherical cow."

I think it's safe to say that Tywin Lannister is not ever going to be up for father of the year, so assume he has the crazy aggression gene and raises his children in the type of home that would lead to the gene being expressed. Though life is much more complicated than Mendelian genetics suggest, Mendel's laws make it easy to estimate the likelihood of passing the gene along if Tywin was a carrier. Joanna was Tywin's cousin and did not seem to have antisocial behavior; however, she is still a Lannister, and I think the crazy runs about as strong in that family as blond hair does. It's safe to assume she's a carrier for the MAOA mutation. Figure 11.3 shows what it would look like if they had kids, where X is no mutation, x is a mutation, and y means it's a boy so there's only one copy of the gene:

Table 11.3

	x	y
X	Xx	Xy
x	xx	xy

The kids turned out as one would expect. I think it's safe to say that Cersei is aggressive, and the Punnett square shows there was a 25% chance of that. The verdict is arguably still out on Jaime, as he seems to have been trained to be aggressive but is naturally nicer than most of the Lannisters—or he is simply better at controlling himself. He had a 25% chance of that as well. Tyrion, we can all agree, does not have a mutated MAOA gene. He also had a 25% chance of turning out normal.

Now, what happens when Jaime and Cersei get together? Table 11.4 would indicate that all of their male children should be affected by the MAOA gene. Because there are no nonmutated MAOA genes coming from Cersei, all of the male children should be unusually aggressive. The female children should all be carriers, but because they cannot inherit an aggressive gene from their dad, they should appear normal. Joffrey's madness was most likely a product of his parents' terrible choices and some unfortunate genetics. The odd kid out here is Tommen. We could argue that he didn't live long enough to truly express his aggression, but he seemed to be relatively nonaggressive. Joffrey seems like an inevitability; Tommen seems like a fluke.

TABLE 11.4

	X	y
x	Xx	xy
x	Xx	xy

So, what about the Targaryens? This is much more complicated since there aren't enough of them living to get a sense of who was a carrier and who was not. Aerys the Mad King clearly had the MAOA gene mutation, but his wife Rhaella has an unknown genotype. She is most likely a carrier or has two copies of the mutated gene. Looking at Aerys's children, Rhaegar seems pretty normal and probably does not have the MAOA gene; the verdict is still out on Dany, but I'm going to guess she got two copies of it. Viserys is clearly aggressive. This leads me to suspect that Rhaella had only one copy of the mutation, making her a carrier. As you can see from the Punnett squares throughout this chapter, madness can be inevitable in just a couple generations. The real question is not how the Mad King happened but why it took so long.

In the end, there's a delicate balance between finding someone similar and finding someone different when you are thinking of having children. If you reproduce with someone who is too different from you, your children could end up with a mismatched set of features that don't work right together. But if you are waaaay too similar to your mate, you can end up with, well . . . Joffrey. Society helps us out with our choices, making marrying your relatives unacceptable and making people that kinda look like you seem attractive. The takeaway here might be something like, "Marry your cousin if you want, but not your aunt." I think this is probably a lesson most of our main characters should learn. I also think I should get a gold star for making it through this entire chapter without making a single *Deliverance* joke.

12

We Do Not Sow

The Science of the Sea

He knew the device it bore: the golden kraken of House Greyjoy, arms writhing and reaching against a black field. The banner streamed from an iron mast, shivering and twisting as the wind gusted, like a bird struggling to take flight.

—Theon Greyjoy, *A Clash of Kings*

Initially, I set out to write this chapter about warfare at sea. I thought it would be fascinating to talk about the battles between the Iron Fleet and the Royal Navy, or about Davos Seaworth and his sons. Two things became clear rather quickly, however: first, I'd already talked about the most spectacular sea battle, the Battle of Blackwater Bay; and second, naval battles without gunpowder aren't super interesting from a scientific perspective. Ships collide with each other, they are boarded by enemy forces, and they are set aflame. Then there's hand-to-hand combat using weapons talked about in other chapters. The addition of cannons to ships required some pretty awesome scientific workarounds, such as figuring out how to balance cannons and counteract recoil on a tippy base, how to aim a cannon when your ship is half on its side, and how not to make a ship so top-heavy it capsizes. But GRRM has made it very clear that there is no gunpowder in Westeros. All that being said, I still think ships are pretty darn cool. Learning how to build ships for a specific purpose, knowing how to sail those ships, figuring out how to stop them from lighting on fire (although no one was super good at that), and figuring out how not to die of scurvy were some pretty spectacular scientific advancements for the time.

As far as ships in Westeros go, I'm going to focus on what's described in the books. The books generally do a good job of keeping things realistic (magic aside), but the show is sometimes a bit off. I mean, if they decided to *cast* a sword instead of *forging* it, how can we expect them to get sail shape right? In the books there are two main types of ships: the longships of the Iron Fleet of House Greyjoy and the war galleys of the Royal Fleet of Stannis Baratheon. This is also anachronistic—war galleys were used in the Middle Ages, and the longship fell out of favor in 1066. The science of a ship's components—the hull and the riggings—is fascinating, as is the method of propulsion. Navigation was an issue until the mid-1700s (with the invention of an accurate and reliable ship's clock), so I'm sure it was something Stannis and crew struggled with. Granted, it was easier in Westeros because one was never very far from land. Still, it could be a problem. What happens if you do get lost out in the Narrow Sea? Scurvy was a real threat to those who had no access to citrus, and it could very well happen to you if you're ever stranded on the high seas. I have written at length about fire, and we saw an entire fleet destroyed by it, but is there any way to protect against it? If there is, I feel bad for Yara (Asha) and the Sand Snakes. (But if I'm being honest, I was kind of relieved when the Sand Snakes left the show. They were much cooler in the book.)

Quick and Light versus Slow and Deadly

As Woody Allen once said, just showing up is half the battle, and the Iron Fleet is known for showing up quickly. It is composed entirely of longships, the Ironborn's goal being to sail somewhere quickly, land close to shore, and buy a lot of things for the iron price. These longships were equipped with both sails and oars, so depending on the weather, the Iron Fleet could be powered by wind or by rowing. This fit their fighting style well. By contrast, the Royal Fleet was developed for more tactical military maneuvers like sieges of port cities. Each fleet's goal is evident in their choice of vessel. The longship was originally developed by the Vikings and used to great effect during the Viking Age (793–1066 AD). In addition to being fast, longships could be sailed through fairly shallow waters. A ship's *draft* is the amount of its hull that needs to be underwater for it to float. Longships were light, so they could draw less water (i.e., have more of the hull above the waterline) and still maintain enough buoyant force to stay afloat. This allowed Ironborn raiding parties to easily land onshore. The downside was that longships couldn't carry much weight without sinking. They were mainly used to transport people from one place to another at high speed, and if you want to be fast, you can't be heavy. By contrast, a war galley was built to carry weapons (and would eventually be the first type of ship to be armed with cannons). They were primarily powered by oars but had sails to use if the weather was good. Generally, a huge crew was required to row such a large and heavy ship. In addition, the ship's draft was much greater, so war galleys were not able to land close to shore. As you can imagine, the difference in goals was reflected in their construction.[1]

There are many forces acting on a ship as it sails through the water, but I'll focus on two: the push forward, whether that be from the wind or oars; and the push back, which is called *drag*. The speed of the ship depends on how much harder the push forward is than the push back. Speed can be increased by optimizing the ship's response to either of these forces. In designing the hull, the goal is to produce the least amount of drag when moving through the water so the ship can move forward without too much resistance. Drag is determined by three parameters: the ship's cross-sectional area as it moves through the water, the roughness or smoothness of the hull, and its shape. When something is moving through a fluid, it needs to push the fluid out of the way. Not all shapes do this very effectively. A cube moving through the water won't push water around the sides as easily as a cone. Shapes that push water out of the way easily are called *hydrodynamic*. The drag created by the shape is called *form drag* because it depends on the form of the object. Michael Phelps is very good at moving his body in a way that minimizes this type of drag. The smaller the cross-sectional area cutting through the water, the easier it is for the water to slip out of the way. Think of walking against the flow of people on a sidewalk during rush hour. You instinctively turn to the side to squeeze through the crowd. A small cross-sectional area going through the water is doing the same thing. You want to shape the hull so that the water flows around it instead of being jammed up against it.

The friction between the ship and the water also causes drag. This is called *skin drag* because it's caused by the drag between the "skin" of the hull and the fluid. Skin drag is typically much lower than form drag, and it's a pretty straightforward idea: slippery objects easily move through water; sandpaper, not so much. The total drag of an object is the sum of its form drag and its skin drag. This total drag force isn't constant; it depends the square of the velocity and the density of the fluid. Michael Phelps's drag force in corn syrup is different than his drag force in chlorinated water. The faster the ship (or Phelps) is going and the higher the fluid density, the greater the drag force. If you are going upstream on a sidewalk, it is much harder to do when there are more people (higher density) and when they, or you, are moving more quickly (increased velocity). Different shapes, even with the same cross section, create varying amounts of drag. A barge has a much higher drag coefficient than a sailboat does. Shapes that help guide water around the hull without creating turbulence have lower drag. An airfoil, or airplane wing, has one of the lowest drag coefficients because the fluid flows smoothly around it. The angled bow shape and small cross section of a longship gives it a relatively low drag coefficient. A war galley hull is also shaped to reduce drag, but it has a much larger cross-sectional area, which, of course, increases drag. A war galley's purpose was to bring siege weapons and troops to a battle. They were essentially the Royal Fleet's tractor-trailers to the Iron Fleet's Ferraris. A Ferrari is fancy and will get you to your destination quickly, but you can't exactly strap a trebuchet to the roof of your sports car.

This is a simplistic view of drag that assumes the water is moving smoothly around the hull, a type of motion called *laminar flow*. Unfortunately, in real life there is turbulence. You've probably heard of turbulence when flying in an airplane, but it can occur in any fluid, water included. In water, you see it as little whirlpools, or *vortices*. Energy, whether from wind or rowing, is used to push the ship forward through the water. Ideally, the minimum possible energy is being used to overcome drag and the rest of the energy generated goes toward sustaining or increasing the ship's speed. Vortices require energy to create; they don't just happen on their own. The energy to create them is coming from the ship's motion. Since energy is neither created or destroyed but can change from one form to other, some of the energy that should be going into pushing the ship forward is now going to vortex creation. Turbulence can dramatically increase the drag, sometimes by as much as 20%. In designing a ship, it's important not only to reduce form and skin drag, which a ship would feel in laminar flow, but also to decrease the turbulence created. Slick hulls do little to decrease skin drag when in laminar flow but can go a long way to reducing the drag from turbulent flow. A rough hull produces a lot of tiny vortices. Rounded shapes also cause less turbulence than sharp edges. I'm a cyclist in my (dwindling) free time, and having an aerodynamic bike and aerodynamic position are very important. There's only so much I can do to change my cross-sectional area, like laying off the Phish Food ice cream, but a bike design can work to decrease the turbulence created as I ride forward. I'm not a pro by a long shot, but even my bike has little tweaks, such as an asymmetric downtube, that decrease turbulence.[2]

Longships were fast not only because of their cross-sectional area, hull shape, and slick sides but also because of their length. Longer ships are faster. Just think of crew racing shells, which are much longer than would be needed to hold 2 people. The reason why is a bit more complicated and a bit less intuitive than drag. When a ship moves through the water it creates a wave at the front called a bow wave.

When moving slowly, this wave has lots of crests and troughs along the ship. Longer waves travel faster than shorter ones, so as the ship moves faster and faster, the wavelength gets longer and longer. At one point it will get so long that the crest is at the bow of the ship, a trough in the middle and other crest at the stern. If the ship goes any faster it will end up with a trough at its stern, which will drop down. Since the stern is wide and flat, this causes extreme drag and slows the ship down significantly. The maximum speed of the ship is determined by how long of a hull wave it can create. Longer ships can have longer hull waves and have a faster top speed.[3]

With its small and fast ships, the Iron Fleet has the advantage in speed, but at a cost. This is where Stannis and the Royal Fleet win the day. War galleys were larger than the Viking longships. To be able to carry a lot of weight without sinking, the ship needs to be very buoyant. I mentioned buoyancy in chapter 4 when I talked about how the army of the dead pulled Zombie Viserion out of the lake, but here is how it works on a ship: For a ship to float, the force down, or the weight of the ship and its cargo, must be less than the buoyant force pushing up on the hull. The force

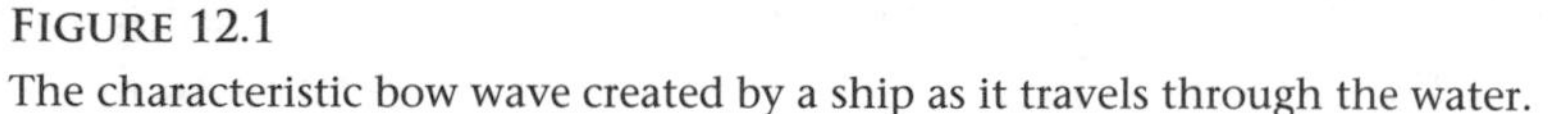

FIGURE 12.1
The characteristic bow wave created by a ship as it travels through the water.

of buoyancy is based on the volume of the ship submerged. The bigger the ship, the more there is to be submerged, and the more it can carry. Stannis's army could arrive armed to the teeth with siege weapons such as ballistae and battering rams. Their size also gave them the advantage in a collision. War galleys were commonly used as battering rams to sink smaller ships. With its large ships and impressive weaponry, the Royal Fleet would have a good shot at taking out the Iron Fleet—but they'd have to catch them first.

Picking Up Speed

Balon Greyjoy and Stannis Baratheon built their navies with different types of ships—one light and quick, one large and made for battle. They also chose different methods of propulsion, with the Iron Fleet relying on wind power and the Royal Navy preferring to row. Although each had the unused method as backup depending on the weather, in general, Yara (or Asha) sailed and Stannis rowed—or, well, their crews did, anyway. Is one better than the other? How do these methods propel a ship? That depends on the shape of things. No, really—it depends on the shape and orientation of the sails, and the shape of the oars.

A longship from the Iron Fleet is primarily powered by sail. The way a ship's sails and masts are set up is called the *rigging*. The type of rigging determines how speedy and efficient wind power can be. Traditionally, a longship had *square rigging*, which is also a popular configuration on ships in pirate movies. Going from the book, one could expect the sails flying the Kraken sigil to be square. In the show, there are two types of riggings: Euron's square-rigged ship with staysails (or side sails) and the much more familiar yacht-style *Bermuda rigging* seen on majority of the fleet's ships. It's odd to see the two mixed, but it is visually stunning, so I'll forgive Weiss and Benioff the inconsistency. (But I'm still never going to forgive them for the cast sword.) The two types of rigging sail by different methods. The square rigging is not too complicated. A mast is set with *yards*, or beams perpendicular to the mast, to which the sails are attached. The yards pivot the sails around the mast, turning them to catch as much wind as possible, like a parachute. Because of their large sail area, they are quite fast when sailing downwind. When the wind is coming from the direction you want your ship to go, however, you are lucky to have oars as a backup. The Bermuda-rigged ship has the ability to sail into the wind—well, sort of. The sails of a Bermuda rig are quite similar to an airplane wing. Hopefully, you remember my explanation of how an airplane flies from a few chapters ago. With an airplane wing, the wind is coming from the front edge and the wing's shape and angle of attack force the air down, providing lift. Take that and turn it on its side and you have a sailboat (figure 12.2). The boat should be pointed into the wind at the optimum angle of attack. As with an airplane wing, the air is forced away from the sail and the boat is propelled in the opposite direction. But it's only the sail of the boat being pushed, so without a counteracting force, the boat would either tip or be pushed sideways across the water's surface. The keel—a wide, flat fin of wood that sticks down from the bottom of the hull into the water—stops this from happening. It is shaped so that

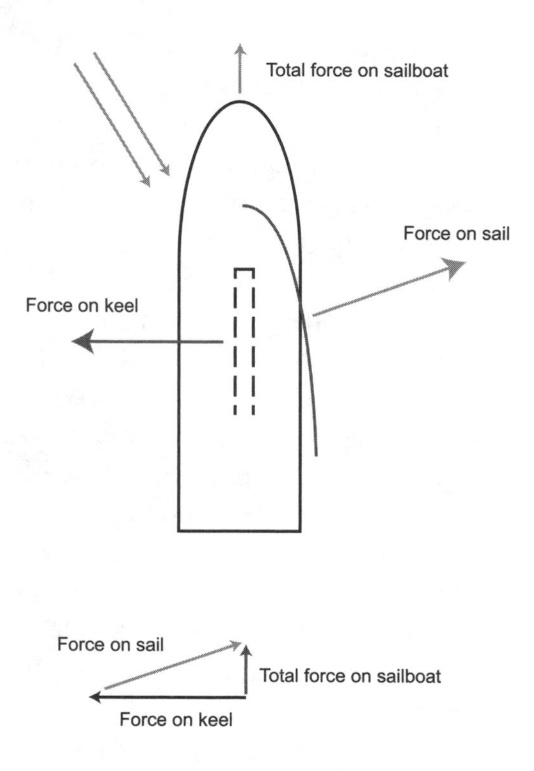

FIGURE 12.2
Force diagram of forcing acting on a sailboat with a Bermuda rigging. Redrafted from a diagram created by James Roche.

it creates little drag moving forward but massive drag if the keel is pulled sideways. This drag stops the boat from being pulled or tipped.

If you want to sail your Bermuda-rigged sailboat or longship into the wind, you can do so by *tacking*, which is zigzagging back and forth at an angle into the wind, so that the boat's ultimate motion averages into a straight line. Unlike a metal airplane wing, a sail can change shape. By moving the sail from one side to other, you can sail into the wind at the correct angle of attack and end up with a net force in different directions.[4]

For his war galleys, Stannis chose a different method of propulsion, one no less dependent on shape: rowing. The physics of rowing is pretty straightforward, so to speak. It's a simple lever action. When the rower pulls back on the handle, the oar blade is pushed against the water. Because every action has an equal and opposite reaction, the ship is pushed forward a little bit. When 100 people are doing this simultaneously in choreographed synchrony, the ship can get going quite quickly.

The shape and size of the oar blade is important in getting the most out of each stroke. The larger the blade, the more power it can generate, and that power combined with the force the rower provides to the lever of the oar is what moves the ship forward. This is a tricky balance to maintain, however—you need someone strong enough to generate that kind of power, and you want the largest oar your rowers can handle efficiently. If there's too much surface area, the ship might move quickly for a short time, but the rowers will tire sooner. It's a marathon, not a sprint.

The shape of the oar is also important in getting the maximal speed out of each stroke. Traditional oars, the type that Stannis's ships would have used, are symmetrical. The blade looks the same on each side of the shaft. In 1991, former Olympic rower Dick Dreissigacker and his brother Pete designed the *cleaver* oar. A main drawback to the symmetrical, or *tulip*, oar was the drag caused by the shaft. With a tulip oar it is impossible to get the entire blade in the water without also dunking in a lot of the shaft. The shaft then drags through the water without providing any real forward motion to the ship. With the cleaver oar, the shaft is attached to the top with the entire blade hanging down. This way the blade can go in the water without much of the shaft having to drag through the water. It seemed like an obvious solution after someone else thought of it. More recently, an Australian rower named Ian Randall developed yet another oar design: the Randall Foil.[5] This oar adds a small lip to the top of a cleaver oar in the direction of oar movement, allowing the oar to make full contact with the water without having to dip the shaft in. In addition, it catches the water that normally flows over the top of the oar causing turbulence. Independent studies found that these oars can increase a boat's speed by around 5%. I bet Gendry's wishing he had a pair of those.

Getting There

If half the battle is indeed showing up, it's pretty important that you show up at the right spot. Navigation at sea can be extremely difficult. We are generally used to navigating by landmarks, but on the open ocean, everything looks the same on a cloudless day. Stars could provide some guidance at night, but nothing like the GPS of today. The ships of Westeros didn't have it too bad since they weren't exactly sailing across the Atlantic and they were rarely too far from land. Still, they needed a decent navigation plan. One of the first methods of navigation was called *dead reckoning*. "Dead" as a descriptor can sometimes imply accuracy or exactness, as in "That's dead useful" or "You're dead-on," so this might sound like a very precise way to measure position. Dead reckoning was anything but. In this method, a ship's position is estimated based on its previous position, how fast you think you're going, and how long you think you've been doing it. There is a whole lot that's not taken into account in this method. In spite of a captain's best efforts, ships don't always go in the direction you think they're going. Oftentimes, a current can cause a ship to drift in one direction or another without anyone noticing. If your measurements are slightly off one time, the error is going to compound until there is a real problem. You can visualize this with pencil and paper: Draw two points far apart. These are your start and end points. Connect them with a line. Now, starting at your start point, draw a second line in the direction of your end point, but angled a tiny bit away. Now observe how the distance between the two lines grows as the lines get longer. At sea, this error can get out of hand very quickly, and with dire consequences. I feel like we've all accidentally played this level on a hike or in the car. We assume we know where we are and make a choice. From there, we think we know where we are and move based on that. Usually, this ends in a fight with your copilot, a promise not to forget the external phone battery next time, and an agreement to ask for directions the next time you stop for snacks. Back then, it ended in scurvy and death.

The astrolabe and, later, the sextant, improved things. If you know the angle between the horizon and an object in the sky, you'll be able to tell your latitude. Going all the way back to chapter 1, you'll remember that the sun appears in a different position in the sky depending on the latitude. The astrolabe was a tool used to measure that angle, comprising a heavy metal disk, usually brass, with an *alidade*, which was basically a moveable pointer. The disk was suspended by a hook on the top so it would hang vertically. The alidade was then moved to point toward a star, the sun, or something else, and the angle was read off the dial. Using this angle and knowing the positions of the sun and stars, it was possible to estimate latitude. It is difficult to hold anything still on a ship's deck, what with all the waves and wind, so this wasn't the most accurate method, but it was a big improvement over dead reckoning. An astrolabe calculated only the latitude of your ship, however—you could figure out whether you were heading north or south, but not east or west. It did nothing to help with longitude.

The sextant was a step up from the astrolabe. It didn't need to be held still and could be used in any direction. The astrolabe had to be hanging vertically for the system to work, so it could never measure angles between heavenly bodies, only between the horizon and the sky. The sextant could measure angles in any direction. The sextant worked with a system of mirrors (figure 12.3). The user would look through the eyepiece

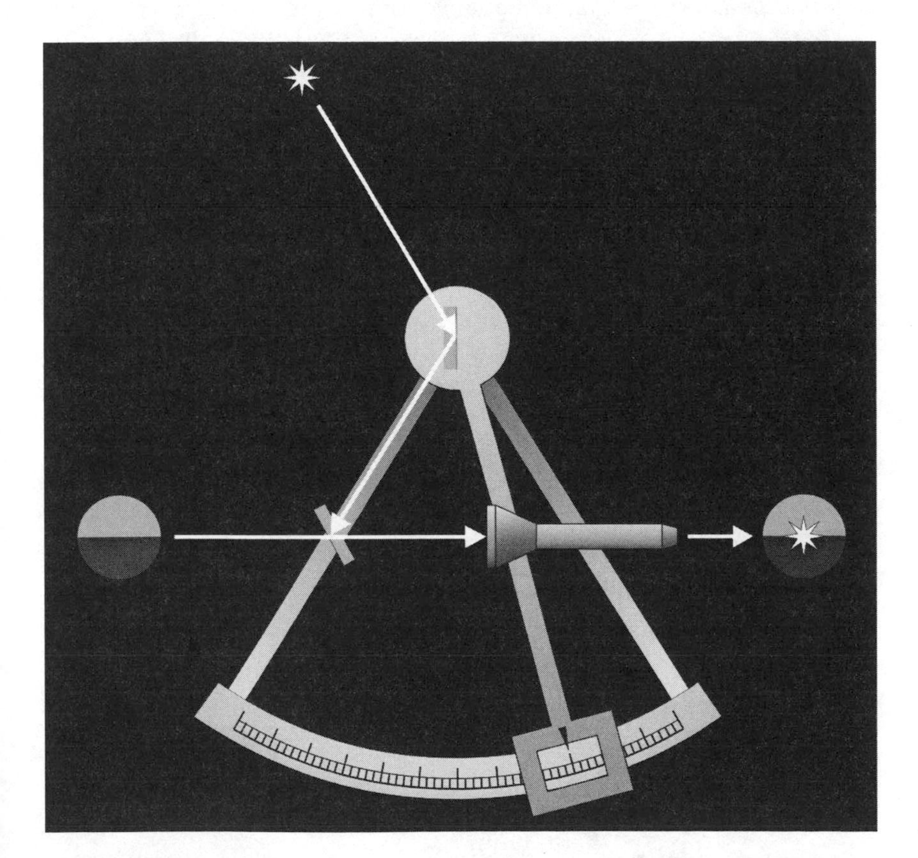

FIGURE 12.3
Diagram of a sextant

to find the horizon and move the arm that tilted the mirror on top of the sextant until the image of the sun or star was reflected onto the image of the horizon. The angle was read and from there the latitude could be determined. The same process is used to measure the angle between two stars. Instead of finding the horizon and reflecting the image of the sun on top of the horizon, you would look at one star and move the arm until the image of the other star was reflected on top of it. Although it was an improvement on the astrolabe, the sextant was still limited to measuring only north and south position, not east and west. Longitude was still out of reach.

One of the best science books I've ever read is about this exact question of how to accurately determine longitude at sea. Dava Sobel's book *Longitude* describes the major challenges of navigation and the quest for a solution. Position involves knowing latitude *and* longitude. Sextants and astrolabes can measure latitude, but longitude is trickier. To measure longitude, you need to know the time difference between where you are and where you started. If I know it's 3 p.m. in Washington, DC, and it's noon where I am (the point when the sun is the highest), and that I'm at 33.44° N latitude, then I know I'm in Phoenix, AZ. The key to solving the problem of navigation lay in finding a very accurate clock. People had very accurate clocks in that day; they just wouldn't work reliably at sea. Could you imagine a pendulum clock working in the middle of the Atlantic? Changes in temperature, barometric pressure, and humidity would also affect the reliability of normal clocks. Sailors were so desperate to measure longitude at sea that in 1714 the British government passed the Longitude Act, which offered incentives of up to £20,000 for the development of a clock that could measure longitude accurately.

John Harrison was able to overcome all these issues and created his "H1" sea clock. Instead of using a normal swinging pendulum, which would be ineffective at sea, he turned the idea on its side. He took two dumbbells and connected them together with springs. They would oscillate back and forth using the springs, not gravity like a normal clock. This removed the problem of swinging pendula. He then created the "grasshopper" escapement to drive these oscillations. A normal pendulum clock's escapement drives the oscillations based on gravity. Since Harrison's H1 couldn't rely on that, he needed to develop a new system. His grasshopper escapement relied on levers and pivots and could be operated at any angle. The resulting clock looked more like a piece of art than a functioning timepiece, but it was successful on shorter voyages. He went on to produce more advanced models, and with the development of the "H4" model he was granted a longitude reward from Parliament in 1773.[6]

Drink Your OJ!

The maesters probably haven't invented an accurate seafaring clock, so I hope that having a sextant and a nearby shoreline keeps Westerosi sailors from getting lost. But what would happen to them if they did? One of the biggest dangers of life at sea is the lack of access to much-needed citrus. Humans, other primates, and (oddly enough) guinea pigs can't survive without vitamin C in their diets. All other animals are able to turn sugar into vitamin C.[7] Somewhere along the line, primates and guinea pigs lost the ability to make the enzyme responsible for that reaction. Since we can't make it, we have to get it from somewhere else: food. This isn't a problem most of the time, since we really don't need much vitamin C and it is usually readily available. There's not a lot of fresh food in the middle of the Atlantic, however, and fish contain only trace amounts of vitamin C. When you deprive your body of vitamin C for an extended period of time, it has no way to make its own, so bad things start to happen. You might think it's possible to "store up" vitamin C for a long voyage by eating a lot of oranges beforehand. This is certainly possible with vitamins D or K, which are fat-soluble, meaning they can be stored in body fat. Unfortunately, vitamin C is water-soluble, and our bodies don't store water the way they store fat. Your body can hold about 1,500 mg of vitamin C and any extra is excreted through urine.[8]

Vitamin C is key to a few different reactions in your body, but the role it plays in collagen production is perhaps the most important one. Remember the discussion about redox reactions in the chapters on steel? That's what vitamin C facilitates in our bodies—it reduces the number of metal ions in enzymes to make sure they can do their jobs. Vitamin C also removes free radicals, which are atoms roaming free in the body looking for something to bond to. If they don't get taken out of the way, they can bind to certain molecules in cells and cause damage. Vitamin C will bind to free radicals and remove them, protecting our cells from damage. It acts as a *cofactor*, meaning a nonprotein participant, in a number of reactions that keep our bodies running. One of these vital processes is the production of collagen, a protein that serves as the backbone of connective tissue, which is any tissue that isn't muscle, skin, or nerve. It is the single most abundant protein in our bodies.

In the 400 years between Columbus's voyage to the "New World" and the invention of steam ships, roughly 2 million people died of scurvy. Scurvy, caused entirely by vitamin C deficiency, begins as general muscle weakness and progresses to gum disease and finally open, bleeding wounds. Vitamin C acts as a cofactor in collagen production by giving hydrogen and oxygen to the amino acids that build the collagen protein, so without vitamin C, collagen production ceases. Collagen is needed to rebuild structures in our bodies. Without it, wounds will stop healing, body parts such as gums and mucus membranes that are often in need of repair will start bleeding because there is no way to repair damage, and eventually weak spots in the skin such as hair follicles will begin to bleed. It takes about three months for a person to blow through their vitamin C stores and start to develop scurvy, and it's a pretty bad way to go. The good news, though, is that it is reversible—it starts to clear up as soon as you get some vitamin C into your body. If you were stuck at sea with a captain who was navigating via dead reckoning, however, you probably wouldn't be near fresh fruit for a while. Worse yet, if you were captured and thrown in a black cell in the Red Keep's dungeons, you were all but guaranteed to die of scurvy—assuming you lasted three months, that is.

Personally, I'm glad I live in an age where I can fly to and from the UK instead of having to sail. Sailing that long doesn't sound fun, and I really like oranges (and having all my teeth). That being said, I'm constantly amazed at the technology people managed to develop centuries ago without the detailed, fundamental knowledge we have now. Even now, many sailors have a sextant and a good watch and know how to use them just in case. From gold nanoparticles in glass and carbon nanotubes strengthening steel to a fire so incredible people are still trying, unsuccessfully, to reproduce it, the Westerosi are pretty badass for a group of people who don't even know why they have seasons.

13

The King's Justice

The Biology of a Gruesome Death

He who passes the sentence should swing the sword.

—Eddard Stark, *A Game of Thrones*

If you have seen the show, heard about the show, caught the show for 30 seconds while changing channels, or even thought of maybe possibly watching the show, you know that *Game of Thrones* is violent. There's no point in dancing around it. Many viewers worry that the violence has shifted from being an integral part of the plot to being straight-up torture porn. Regardless of your feelings on this point, if you are reading this chapter, you are fully aware of how many unique and painful ways *Game of Thrones* can come up with to kill a person. If you are reading this chapter, please understand that I'm assuming you are interested in learning more about the reality of this fictional brutality and that there will be no holding back. It's gonna get rough. After researching and writing this chapter, I can tell you in no uncertain terms that the reality of these deaths has caused me more nightmares than all the shows and books combined—and that is saying something. There is something about contemplating the biological and physical reality of death that puts one in the condemned's shoes (or noose) in a way a work of fiction never could. So, if you are ready for that, read on. Know that however bad this chapter makes you feel as a reader, it was way worse to write. Writing this chapter during long-haul flights has also gotten me more than a few questionable looks. If you want to make sure your seatmate won't talk to you on a long flight, I highly recommend having a Google doc with a subsection titled "Beheading" open on your laptop.

Now, if you are looking for instructions on how to be resurrected after being stabbed, I'm sorry to say that this chapter will be of no help. Personally, I'm Team Lord Commander all the way, but as a scientist there's not much I can do. (I suggest you ask the Red Woman.) You have been warned.

BEHEADING

There has been a lot of research into what exactly happens at the moment of death. Does the brain shut off the moment the heart does? Does it take some time for the brain to get the message that the heart has stopped? What can near-death experiences and the movie *Flatliners* really tell us about this? Researching the biology of a beheading is an excellent place to start exploring these questions that you probably didn't know you had until now. In a beheading, the head is severed from the body by a blade of some sort. The questions I'm hoping to answer in this section are: How does the method of beheading change what is experienced by the person executed, and would you be aware of your own beheading after your head has been separated from your body? Anne Boleyn famously asked Henry VIII for a French executioner, who would use a sword instead of the traditional English axe. Was this worth it as a dying request? During the French Revolution, many people were said to have made facial movements after their head had left their body. Scientists even asked condemned prisoners if they would try and communicate after their decapitation. Were the movements just imagined by observers or were these people really able to comprehend the crowd cheering their death? How painful would this type of death be? The most traditional methods of beheading are the sword and the axe. Indeed, these are the two seen in *Game of Thrones*. It was believed that the sword was less painful and more effective than the axe. Considering their relative sharpness this would make some sense. It's a very interesting physics problem and one better solved by looking at a method of beheading not shown in *Game of Thrones*: the guillotine.

Probably the most (in)famous method of beheading was the guillotine, developed by Dr. Joseph-Ignace Guillotin and harpsichord maker Tobias Schmidt. Ironically, Guillotin first proposed the contraption that bears his name in a bid to end the death penalty altogether. He felt that the first step in doing away with it outright was to figure out how to administer it quickly and painlessly. It didn't work. The guillotine was in use in France from 1792 until 1977. To give you some historical context, the last judicial execution by guillotine occurred a few months after the first *Star Wars* movie was released, the year that Atari hit the shelves, and the year you could buy your first Apple II computer. Jimmy Carter was president. It was not that long ago. When thinking about how painful it might be to be beheaded, I was curious about the likelihood of a headsman doing it in one blow. Would it be like Ned's execution of the Night's Watch deserter, or more like Theon's botched beheading of Ser Rodrik Cassel? It is complicated to figure out the force needed to cut through a person's neck since there isn't a lot of useful data on the subject. This is why the guillotine might be a good place to start. Considering it was used about 30,000 times and was deemed to be very effective, it's safe to assume that the force it provided was enough to comfortably, reliably, and repeatedly sever a head in one blow. As I'll explain more fully in the section on hanging, tackling this problem head-on (get it? Head-on?) is easiest by looking at energy. Once we find the energy per unit area provided by the guillotine, we can make a pretty good estimate of how much force a headsman needs to apply.

When the guillotine blade is at the top of the structure, it's about 2.2 m high. The blade itself weighs about 7 kg, which is not very heavy—it's about the weight of the average Thanksgiving turkey. To add more heft to the blade and increase the energy, it would have when slicing a neck, a *mouton*, or weight, was added to the top. The whole contraption ended up weighing about 37 kg, or a little less than me without my head (OK, more like 50% of me). At the top of the guillotine, the blade has about 820 kJ of potential energy. Some of that is lost to friction on the way down and some is lost to the friction of cutting through the neck. The blade was angled to reduce the amount of friction experienced at one time so as not to slow the blade down too quickly; and if you've ever watched French chefs, you know they like slicing with their blade at an angle. When it comes to headsmen, however, it's less about the amount of force applied by the executioner than their skill with an axe. The axe used by the English could be sharpened to the same thickness as a sword, so, assuming the executioner is strong enough, the axe should be more humane since it is heavier. An axe provides more energy per unit area than a sword. Unluckily for many Brits, the axe is much harder to wield and the headsman much less practiced. The average neck is narrow, but the back and the head are not. The executioner must land a blow on the narrowest part of the neck to have enough force to make it through the whole thing. As a blade moves through the neck, energy is lost to friction and mechanically breaking bonds. Once the potential energy is used up, the axe stops. If it hasn't made it through the neck, the prisoner is in for a much worse day. Just as hanging was a practiced craft, so was beheading. The execution method of choice in England was hanging; beheading was reserved for the highborn. Hangmen would often moonlight as headsmen. The goal of hanging, however, was to keep the head firmly attached to the body, so a hangman did not always make a good headsman. It often took them three or more strokes to fully decapitate someone. In France, however, beheading was the preferred method of execution. Their swordsmen were able to aim correctly and sever heads in one blow.

What happens once the executioner finally severs a head? What is going on up there in the gray matter? For a brain to be conscious, it has to have enough oxygen for the neurons to send electrical impulses. I'll go more into how neurons work and how to mess with them in the section on poisoning, but the key point is that when neurons are able to fire, a person may very well be aware of what's going on. Some "experiments" were done during the French Revolution in which the condemned were asked to communicate with a scientist in the crowd after the fall of the blade. Many onlookers swore they saw signs of recognition on the disembodied faces. This was—and continues to be—a hotly debated topic. Ischemia is the technical term for loss of blood flow to the brain. Without constant blood flow, the potassium, sodium, and calcium channels cause excessive amounts of potassium to linger in the intercellular space between neurons and cause the neurons to retain too much potassium and chloride. In a human, the magic number for ischemia has long been thought to be 10 seconds; after that point, unconsciousness sets in.[1] To know what that would be like, set a stopwatch for 10 seconds. Hit go and see how much you take in and think about in those 10 seconds. Kinda scary to think you might be conscious for that long after your head had been removed from your body. The time from loss of blood flow to the brain to unconsciousness was long theorized but not tested with modern equipment. (Imagine trying to get that past the institutional review board.) A group from the Netherlands decided to use their equipment and some lab rats to settle the debate about consciousness after beheading.[2] Beheading via cervical spine dissection was long used as a way to euthanize rats no longer needed in experimentation, and was preferred over pharmacological euthanasia for experiments in which the brain needed to be studied. There was much debate as to whether or not this was a humane form of euthanasia, however. To test this, Clementina van Rijn hooked electrodes up to rats' heads and measured the duration of significant electrical activity after the rats' heads were severed. This was a surprisingly difficult paper for me to read; I had to take a break after contemplating the idea that the university machine shop had to build a rat-sized guillotine. They used the tiny terror to systematically cut off the rats' heads and then measured the electrical signals in the brain. They found that it took about 2.7 seconds for brain activity to slow to a point that would indicate unconsciousness. During this time, the disembodied head seemed to be making a chewing motion. Scaling up to a human head, they estimate that someone would have about 7–10 seconds before they would lose consciousness. In discussing this section with others, I regularly heard people say, "If it was me, I'd probably be screaming," or ask, "Did the heads in the French Revolution ever scream?" The answer is no, they did not. To make a noise, air from the lungs must pass through the vocal cords and out of the mouth. Once a head is removed, the lungs and the mouth are no longer connected, so it's not possible to scream. I'll leave it up to you to read the rest of this chapter and decide how scary death by beheading may be. After you read about burning at the stake, I think you would choose the sword.

A Golden Crown

Viserys's death was certainly one of the most fitting in the series: in his dying moments, he finally got his golden crown. The idea of killing someone with molten metal has historical precedent; the Spanish Inquisition's prosecutors were particularly fond of pouring molten metal down people's throats. Before I get too far into this discussion, however,

I need to address one glaring physics error in Viserys's death scene. The melting point of gold is about 1064°C. The hottest a wood fire can get is about 585°C. Those who have read the section on dragon fire might have already figured this out. In reality, Viserys would more likely have died from being knocked out by the blow of the heavy gold being dumped on his head. I know this is a huge error, and in keeping with the theme of this book, I should stop here; however, I'm too interested in what would happen if gold were poured over someone's head to stop now. I don't think anyone would argue that molten gold dumped on your head will kill you, but the question is how exactly it would do that. What, exactly, would it say on Viserys's death certificate? The options are suffocation due to his mouth and nose being blocked by gold, shock from having molten gold poured on his head, or the cooking of his brains. Since I think the most likely cause of death is brain cooking, I'm going to start by looking at how long it might take to boil a brain and see how that compares to other two possible causes of death.

I'm not the first to be interested in this topic (obviously). In 2003, a group of pathologists in Amsterdam approached this question experimentally by taking a cow larynx (not one attached to a cow; rather, one that a dead cow was no longer using), covering one end with tissue, and pouring molten lead at a temperature of 450°C down the other end. The steam produced immediately blew out the tissue paper at the other end.[3] Seeing as steam seems to be the most pressing issue, I'm going to look at how long it might take for a brain to boil if it is heated by molten gold poured over the skull. To answer this there are two things to look at: the energy required to boil the water in the brain and how fast that energy can be transferred through the skull. The brain weighs roughly 1.4 kg and comprises 73% water, which means the average brain holds about 1 kg of water. This makes the math very easy. There are two factors involved in boiling: First, the temperature needed to raise the substance to its boiling point. In the case of water that is 100°C. Then, energy is needed for the actual boiling. It takes a lot of energy for a substance to change physical state; in this case, from liquid to gas. (Chapter 2 goes into more detail about the physics of this as it relates to the Wall.) The unit of energy is the joule, but for this example it's a bit more practical to use kilojoules (kJ). It takes 2260 kJ to change a kilogram (kg) of water to steam. To find out how much energy is needed to raise the temperature of 1 kg of water by 63°C (body temperature is 37°C) the equation to use is the specific heat formula,

$$Q = cm\Delta T,$$

where Q is energy in kilojoules, m is the mass in kilograms, ΔT is the change in temperature, and c is a constant that's different for each material and indicates how easy it is to heat something up. Plugging everything in, we get that it takes roughly 268 kJ to warm the brain up to the boiling point. Adding the two numbers, we get that it takes about 2528 kJ to boil all the water in the human brain.

To figure out how long it would take to boil the brain, we need to determine how quickly heat can be transferred from the molten gold through the skull to the brain. This is a *very* rudimentary look at this calculation. Heat transfer from the outside of the brain to the inside of the brain will be different from the transfer of heat from the gold through the skull, and there will some additional time associated with heat moving from the outside of the brain to the inside, so this number will only be a really rough estimate. However, it will give us enough of a handle on the time frame to see if boiling the brain will take less time than, say, suffocation. The equation that determines how quickly heat transfers through something like bone is pretty similar to the equation that tells us how quickly something heats up:

$$\frac{Q}{t} = \frac{kA(\Delta T)}{d},$$

where Q/t is the energy delivered per second; k is a constant that indicates how quickly energy moves through a given material (in this case, bone); A is the surface area in square meters (in this case, the surface area of the skull); ΔT is the difference in temperature between the two materials (in this case, 1027°C, the difference between the temperature of molten gold and body temperature); and d is the thickness through which the heat is traveling in meters. The thickness of the average male skull is 7.1 mm, or 0.007 m.[4] The surface area of the skull is 0.98 m^2, according to a group that measured many different skulls.[5] The constant, k, was frighteningly easy to find and ranges between 0.410 and 0.630, so I'll assume 0.5. Taken together, we find that bone transfers about 718 kJ/s to the brain. Looking back at how much energy is needed to make a brain boil, we can estimate that it takes roughly 3.5 seconds to fully boil a human male brain.

Again, this is really just an order of magnitude estimation of the problem and not a precise number, but it's clear we're talking seconds and not minutes. It would take about 3 minutes for Viserys to suffocate with a nose and mouth full of gold and much longer to die from shock. He would be dead before his whole brain boiled for sure, so 3.5 seconds is really the upper bound on death by brain cooking. Although this isn't an exact figure, I think it's safe to say that Viserys died pretty quickly, he probably didn't have time to jerk around the way he did in the book, and he most likely did not suffer the way most fans wish he had.

HANGING

It may not be one of the flashier execution methods used on *Game of Thrones*, but hanging seems to be the most ubiquitous. The first recorded execution by hanging was in Homer's *Odyssey* and was carried out via suspension hanging. The Greeks may have invented hanging, but the British perfected it. It is one of the oldest ways of executing someone, second only to beheading as the longest-used judicial method of execution. Unlike beheading, it is efficient in that many people can be killed at once with a single executioner. Hangings were a joyous town event and a reminder to citizens of what would await them if they broke the law. Public executions were not just a method of offing criminals but a way to deter others from committing crimes. Hanging people, however, can be a tad more difficult than lopping off their heads. So how does hanging really kill someone? Is their neck broken? Do they suffocate? Is the blood supply to their brain cut off? Turns out the answer to all these questions is "It depends." There are several different methods of hanging—long drop, short drop, and suspension—and they all lead to death in different ways. As far as I can tell, only two of the three are used in the book or show, but I'd like to talk about all of them, so I will.

Suspension hanging is possibly the oldest method of hanging. In this method, the person to be hanged is suspended by their neck until the heart stops. Generally, the noose is placed around the person's neck and they pulled up till their feet are off the ground. If you are a book reader, you know that Brienne of Tarth can explain this method all too well. After her encounter with Lady Stoneheart, she said that nothing had ever hurt so much. This method is no longer used (as far as I've been able to find) as a method of execution now that the long-drop method is standard. It would seem that the cause of death in this kind of hanging is pretty clear—you can't breathe if there's a noose around your neck. It's more complicated than that, however. You have about a 40% chance of dying by suffocation and a 60% chance of dying from the constriction of either your carotid artery or jugular vein. So how you die depends on the placement of the noose's knot. If the knot is on the left or right side of your head, the pressure will be on your arteries, veins, or both rather than on your windpipe. The carotid arteries and the jugular veins are squished, which stops blood flow to and from your brain. It is much, much quicker to die this way. It only takes about 4.4 pounds of force to compress the jugular vein and 11 pounds of force to compress the carotid arteries, but 33 pounds of force to compress the trachea. If the arteries or veins are cut off, blood flow to your head stops and you would pass out within 15 seconds or so. If you are into forensics, it's possible to determine whether the jugular vein was compressed more than the carotid artery. If the jugular is more compressed, the blood can get into the head but can't leave. Little capillaries burst from the pressure, leaving telltale red spots called *petechial hemorrhaging*. You might know this term from such shows as *CSI*, *NCIS*, and *Law and Order*. Facial petechial hemorrhaging is indeed a marker of strangulation for just this reason, but contrary to Hollywood's presentation it can occur on any part of the head and face, not just the eyes. The absence of petechial hemorrhaging doesn't rule out strangulation; it just means that if the person were strangled, the perpetrator pushed more on the arteries than the veins. After passing out, the condemned would eventually die from either suffocation that they wouldn't feel, or from blood not circulating through the brain.

If they are one of the unlucky 40%, as I believe Brienne was, the pressure is mainly on the trachea and death is caused by asphyxiation alone. This usually happens when the knot is at the back of the head. Death by suffocation takes much, much longer and for the majority of it, the one being hanged is conscious. In these cases, there is often damage to the hyoid bone or larynx. It is presumed to be quite painful both because pushing that hard on the trachea would not be comfortable and because struggling to breathe is a terrible experience. Just try holding your breath for too long.[6] This type of hanging might seem unlucky, but back in 17th and 18th centuries (and, one assumes, in Westeros), having the knot in the back might be your get-out-of-death-free card. The tradition with suspension hanging was to hang people for about 30 minutes. In more than a few cases, the pressure on the trachea was just enough to cut off air and cause unconsciousness, but not enough to kill. The hangman was even known in many cases to pull on the feet or sit on the shoulders of the unlucky soul on the end of the rope to make sure his work was done. In spite of all this, in many cases those thought to be dead would wake up in transit or on the autopsy table. It became so common that friends and family of the "dead" would try and revive them. Luckily, in many cases, if you were revived it was considered a miracle and you were set free. The case of "Half-Hangit Maggie" is particularly interesting.[7]

Short-drop hanging is a bridge between suspension and the Brit-perfected long-drop method. It was the first attempt to make hanging more humane and also less prone to failure. The manner of death in a short-drop hanging is very similar to that of suspension hanging, in most cases—the person dies from vein and artery compression, suffocation, or both. However, if everything is positioned correctly and the short drop is a little longer than usual, it's possible for enough force to be applied to break the neck and cause almost instantaneous death. Death from short-drop hanging is sort of a crapshoot, but it's much more likely to ensure death. This method uses a 1–1.5-foot-long rope. The condemned is put on some sort of platform that is quickly removed, usually a chair or a cart or, if you are in the Old West, a horse. Jon Snow used this method to hang the Brothers who stabbed him. Short-drop hanging is only slightly better than the suspension method in that there is a small chance death is instantaneous. From a scientific perspective, there's not much to be said about it other than that it inspired the hangmen of Britain to find a way to ensure the quick and painless death that sometimes occurred in a short-drop hanging.

The long-drop hanging method is not explicitly seen in *Game of Thrones* because it didn't become the standard method until fairly recently, around 1872. It was seen as a more humane method of execution yet was still showy enough to deter would-be criminals. Although this isn't the type of hanging seen in the show, it is usually what comes to mind when one thinks of judicial hanging, and, from a physics point of view, it is by far the most interesting. The goal of this method is to snap the neck as quickly as possible to induce death quickly and painlessly. When the neck is snapped the spinal cord is severed and the person being hanged dies quickly. This is not always easy to do, so hangmen, and British ones in particular, were seen as artists whose goal was to cause the least amount of pain to their victims. William Marwood was the first to begin thinking about how a longer drop could cause a less painful death. He recommended a drop of 7–10 feet, much longer than the short-drop method. There were a lot of disagreements about the right way to make the long drop work, from the drop distance to the placement of the knot. Whether or not (knot?) they knew it, hangmen were really experimental physicists.

Traditionally, each hangman had his own method; regardless, the key thing to know in long-drop hanging is the energy required to break a human neck. By knowing how much energy needs to be applied, you can figure out how far the person needs to drop to end up with energy to break the neck. (I'm required to say here that energy is neither created nor destroyed; it just changes forms. Sorry, too much physics training.) In your physics class, energy is usually measured in joules, but British hangmen preferred the unit of ft · lbs. When the prisoner is standing on top of the trap door with a noose around their neck, they have energy of position, or potential energy (much like the potential energy of the suspended guillotine blade). How much potential energy the prisoner has depends on the height of the platform, or rather how far the fall will be. Potential energy can be difficult to wrap one's head around because "energy" usually implies action, and someone standing on the gallows isn't doing much else besides sweating. The condemned had to walk up to the gallows, however, and thus had to provide the energy sufficient to cause their own death. In addition, the prisoner's high position contributes potential energy that could turn into energy of motion, or *kinetic energy*. On the ground, it's not really possible to move by falling, but getting down from a high platform requires movement. As the condemned falls through the trap door, that potential energy changes into kinetic energy. When the prisoner's neck hits the end of the rope, all that kinetic energy goes into the torque created by the angled knot to snap the neck. The key point is that all the energy the prisoner started with—the potential energy—is eventually transferred to the prisoner's neck. To make sure there is enough energy to break the neck, we need to know two things: how much energy will break a neck, and the height required to transfer that energy into the neck after the fall. Potential energy is $U = \text{weight} \cdot \text{height}$, where U is the universal symbol for potential energy for reasons I've never understood. There have been varying estimates over the years of how much energy it takes to break a neck. In the 19th century, executioners estimated it to be about 2,240 ft · lbs, but this figure has been revised over time. A 1947 US Army manual outlining military execution procedures states that it would take approximately 1200–1400 ft · lbs to break a neck. They also specified that a military band should be present to play a lively tune after the execution.[8] I don't know how accurate their numbers are, but you can't fault their style. From here, determining the length of the rope is pretty straightforward. Divide the neck-breaking energy by the weight of the prisoner and you get the height of the drop needed and thus the length of rope that should be used.

At the dawn of long-drop hanging in the 19th century, people understood the physics of all this, but they were just making educated guesses based on past hangings about the amount of force required to break a neck. The two premier British noosers (which is a fabulous Scrabble word, FYI) were William Marwood and Albert Pierrepoint, and each had his own table of drops.[9] The 1947 Army manual has yet another. You might be thinking, "Hey, whatever, err on the side of longer, really break the neck, and we're all good." Unfortunately, there is an upper bound on the force you can apply. Apply too much force at the end of the rope, and the condemned's head will pop right off. It is a truly gruesome event and one that hangmen hoped to avoid. It is still unclear how much is too much and how little is too little. Luckily, this is not as much of a pressing matter these days. If you are interested in a more detailed history of hanging, Mahmoud Rayes, Monika Mittal, Setti S. Rengachary, and Sandeep Mittal published a paper in 2011 that offers a fascinating romp through the lives of hangmen throughout the ages. I can't say I recommend it, but it is very informative.[10]

There is one type of death that hasn't happened yet but is worth talking about: manual strangulation. To date, no one in the show or books has actually been manually strangled by someone else, but book readers know that eventually Cersei Lannister will get what she deserves, according to the prophecy of Maggy the Frog: "And when your tears have drowned you, the valonqar shall wrap his hands about your pale white throat and choke the life from you." From a science perspective she'll pretty much die the same way as someone in a suspension hanging, only most certainly from cutting off blood supply to the brain. It takes about 33 pounds of force to crush the trachea, and quite frankly I don't think either of her younger brothers would have that strength in their hand(s). I do think, however, that one of them could muster the 4.4 pounds of force to compress her veins, or the 11 pounds of force to compress her arteries. In manual strangulation cases, petechial hemorrhaging is almost always present because of the difference in force. Because it only takes half the force to completely compress a vein, it is much more likely the veins are damaged more than the arteries, which leads to bursting capillaries. In all likelihood, Cersei's pale white throat would end up a bit spotted.

POISON

Poison has long been seen as a weapon of the cold and calculating. Ned Stark called it the "woman's weapon," and although Oberyn Martell may disagree, the majority of successful poisonings in *Game of Thrones* have been carried out by women. (To be fair, Arya Stark used poison out of convenience—it was much easier to off all the Freys at once with poison than with repeated throat-slitting.) Westerosi poisons seem to come in a variety of types, many similar to real-life poisons, but in several cases, the method of death may have been played up a bit for the cameras. The poison used at "The Purple Wedding" has effects far beyond what a real poison could accomplish, but "The Pale and Sleepy Wedding" doesn't have quite the same ring to it. Poison also has the advantage of being hard to detect and its effects often mirror those of a sudden illness. I'll talk specifically about poisons that mirror those in *Game of Thrones*, so I'm skipping over some of the common ones such as cyanide (and, in particular, its place in history), although if I had infinite space they would be fun to talk about. In figuring out what types of poisons might have played a part in deaths on *Game of Thrones*, availability also needs to be taken into account. The science of poisoning could easily be extended into its own chapter, so please forgive me if I leave out your favorites.

The proverbial kickoff to the great *Game of Thrones* occurred with the poisoning of Jon Arryn. Things went downhill fast for the Starks after they received a letter from his wife, Lysa, about the unusual circumstances surrounding his death. Lord Varys offered his suspicions that the cause of Arryn's death may have been the Tears of Lys, an untraceable poison that mimics death by intestinal ailments and is, according to Oberyn Martell, a "favorite tool for impatient heirs." True to the assumptions of a poisoner's gender, it was Jon's wife Lysa who ultimately did him in. From a scientific point of view, what might the deadly Tears of Lys be? Is there such a poison? Arsenic, nicknamed "the inheritance powder," is a commonly used poison, and the symptoms of arsenic poisoning mimic those of other diseases such as the stomach flu. The only seeming difference between the two poisons is that Tears of Lys is supposedly very rare and expensive, while arsenic is pretty easy to find. Arsenic is not a compound but rather an element—number 33 on the periodic table, to be exact. It's classified as a *metalloid*, which means it has some properties of metals and some properties of nonmetals, and it is the 53rd most abundant element in Earth's crust. It is a naturally occurring mineral and is found in both soil and rocks as well as mixed in with other minerals. Because it's found in soil, it can also find its way into unfortunate places such as groundwater and plant roots, poisoning wells and ending up in crops. Arsenic in its elemental form is not very good for you, but it's not nearly as toxic as it is in its compound form, arsenic trioxide. Known as *white arsenic*, this is the stuff that is used to off one's friends and family. Unlike the fictional Tears of Lys, it's quite easy to come by. Arsenic trioxide is created when anything that contains arsenic, such as coal and gold- or copper-containing ores, is heated. During the combustion process, two arsenic atoms bond with three oxygen atoms to form a white powder. Historically, this was used in everything from rat poison to teething medicine. In the 19th century, you could walk into any drugstore and buy a bottle. Arsenic wreaks havoc on a cell's ability to make and use energy and stops the work of a crucial enzyme in cellular respiration. If that wasn't enough, it inhibits thiamine (vitamin B1), raises the production of hydrogen peroxide, and messes with a cell's potassium channels. In short, it interrupts the chemical pathways that make usable energy for cells. Without energy, the cells die off, eventually killing the unfortunate victim.[11]

One of the most surprising aspects of the case of Jon Arryn was the certainty with which Varys told Ned that Arryn had been poisoned. How could he have known? We can assume that the spectroscopy technology used to detect arsenic in a blood sample is not available in Westeros, but that doesn't mean the poison would be untraceable. The Westerosi may not have access to modern, high-tech CSI devices, but they can get by without it because arsenic is one poison that can be detected without the aid of sophisticated lab equipment. In fact, it was one of the first poisons to be easily detected in a corpse. Arsenic causes telltale stomach lesions and oftentimes leaves a crystalline coating of poison in the digestive tract, both of which are easily seen in an autopsy. We've seen that maesters are adept at autopsies, so if Arryn's death was at all suspicious, the maesters could have easily verified their assumptions. So, the obvious question might be, "How did Varys know?" but the better question is, "Why did the maester tell him and *only* him?"

In the (sort-of?) death of the Mountain at the hand of Oberyn Martell, he fell victim to a poison-tipped spear. Technically, the poisoner wasn't a woman, but I think this might not count since weapons and fights to the death were also involved. We'll get to the physics of Oberyn's death in the next section, but the Mountain's death (I am not going to get into the biology of resurrection) was painful and slow. His skin festered, he was paralyzed, and worst of all, he was conscious for the entire process. Unlike the ingestible or absorbable poisons, Oberyn's "manticore venom" seems to require injection right into the bloodstream.

Poison-tipped spears and arrows have a long history, dating back to Homer's *Odyssey* and Virgil's *Aeneid*, and they are still used as weapons in South America, Africa, and Asia. The term *toxin* comes directly from the Greek word for "bow," *toxon*. A poisoned arrow, dart, or spear may seem like an excellent weapon; however, the likelihood of accidentally puncturing your own skin with your poisoned weapon was high enough to deter some from using them. The traditional poisons used were all found in nature and were either plant- or animal-based. The components of the poison depended on what was available. Everything from monkshood to poison dart frogs has been used to make poison for use on weapons. The fictitious manticore is described as having the head of a man, the body of a lion, and the tail (or stinger) of a scorpion. So, in trying to figure out what type of poison was used by Oberyn and what kind of death the Mountain suffered, scorpions would be one place to start. A scorpion sting is often compared to a bee sting. I've never been stung by a scorpion, but I have friends who have been, and they say this is a gross understatement. It hurts! There are many scorpion species, but the deadliest is the deathstalker scorpion; luckily, these are not native to the United States. Although the sting of one is not usually enough to kill an adult, a concentrated amount on the end of a spear certainly could. Its venom is a powerful mix of neurotoxins. The most active, chlorotoxin, works by blocking a certain type of chloride channel, CLCN1, in muscle cells. The proteins in the toxin are the perfect shape to plug up the channel and stop chloride from getting in. Chloride is essential to making a muscle cell reactive to the electrical impulses from nerve cells. When chloride can't move through the channels, paralysis occurs. This is one of the Mountain's main problems. This is the only effect of deathstalker scorpion venom, however; it doesn't cause the necrosis seen in the Mountain's wound. For that, we need to look elsewhere.

Manticore venom also seems to have an effect similar to that of brown recluse spider venom. The brown recluse (*Loxosceles reclusa*) is one of the deadliest spiders on Earth, and its venom causes skin necrosis much like that seen in the Mountain's wound. The spider's venom interacts with the membranes of the cells it comes in contact with and destroys them, killing the cell. A cell membrane is made up of two layers of molecules called *phospholipids*. They are little heads with two tails. In the bilayer, the tails are in the middle with the heads on the outside. Brown recluse venom contains a protein that rips the heads off the phospholipids, destroying the cell membrane and killing the cell. This causes the necrosis that radiates out from the wound. The spider's venom, however, works over a number of hours, not the short time scale we saw on screen.[12] If I had to make a guess at what GRRM's version of a manticore might be, based on its venom alone I'd say it was the love child of a deathstalker scorpion and a brown recluse with a fast-acting venom mutation.

The Purple Wedding was one of the most wonderful yet horrible scenes in season 4. Joffrey finally got what he deserved, and it was pretty graphic—the "Strangler" quite literally chokes its victims to death, leaving a purple-faced corpse in its wake (hence the nuptials' colorful moniker). So, what exactly did Olenna Tyrell slip into his wine goblet? What kind of poison is the Strangler? It has to be available in crystal form since it came from Sansa's necklace, which means it's probably tasteless and easily dissolved in liquid. The most likely candidate is strychnine. It is available in crystal form and kills by causing muscle contractions and eventually asphyxiation. The poison is derived from the fruit or seeds of plants belonging to the genus *Strychnos*, so it would certainly be something the maesters could find in nature. It doesn't take much to kill; a fatal dose is only about 16 mg/kg of body weight, and strychnine can kill in as little as 15 minutes at high doses. If you assume that Joffrey was about 45 kg, which is the average weight of a 13-year-old kid, it would only take about 720 mg to kill him. (For comparison, the average dose of ibuprofen is 400 mg.) Signals in the central nervous system are carried by molecules called *neurotransmitters*. Strychnine binds to the same receptors as glycine, an inhibitory neurotransmitter, slowing down the electrical signals that control the brain. When strychnine fills in for glycine, the electrical signals go into overdrive causing muscle contractions and eventually death by suffocation. This seems like the most likely poison responsible for Joffrey's death; however, it works on all muscles of the body, not just the throat. Ingested poisons act on the entire body, rather than on a specific spot, as an injected poison might. So, although the Strangler's effect on Joffrey certainly made for a dramatic death scene, there's no poison that could have caused just the throat to close up.

In one of the rare cases of a man using poison in *Game of Thrones*, Jaime Lannister gave Lady Olenna the gift of a supposedly quick and painless death. It wasn't fast enough, however, to stop her from twisting the knife one last time. There are two classes of drugs that could be responsible for such a quick(ish) and painless death: barbiturates and opioids. In modern times, these classes of drugs are used extensively for medical purposes:

barbiturates as anesthetics and opioids as pain relievers. As most people have guessed, Westeros's milk of the poppy is most likely an opioid derived from poppies. To control pain response, our bodies secrete natural opioids that bind to special opioid receptors in the central nervous system. When the natural opioids bind to the opioid receptors, it sends a signal to the brain to block pain and slow breathing. We don't walk around in a constant state of opioid stupor because our bodies synthesize very small amounts of these opioids. Most of the receptors don't have an opioid buddy at any given time, so we aren't too numb to pain. Add in an opioid of some sort, and those opioid receptors fill up and tell our bodies to calm down and block pain signals. Unfortunately, opioids that come from an external source never bind quite right to the receptors, so in addition to calming and controlling pain, they send dopamine synthesis into overdrive. This is why opioids are so addictive: they trigger the brain's reward system. In an overdose, the calming effect is too strong and the respiratory system relaxes so much that it stops. The tricky thing with trying to kill someone with opioids is that it is quite difficult to find a dosage that will reliably kill. Since Jaime very much wanted to leave the castle knowing Lady Olenna would be dead, I doubt he would have chosen milk of the poppy. In addition, it would take a fair amount to kill her, and Jaime had only a very small vial.

It's more likely the poison came from the barbiturate family. The sleep aid sweetsleep seems to be the most likely candidate for a fictional barbiturate, because a very small amount is said to induce sleep and a small amount can cause a sleep from which you won't wake up. The first barbiturate was synthesized in 1864 by Adolf von Baeyer. He combined concentrated urea with a compound derived from the acid of apples. Even though the first barbiturate was derived long after the supposed time period of *Game of Thrones*, there's no reason the maesters couldn't have found something similar; they had access to both urea and apples. Barbiturates depress the central nervous system, making you groggy, disoriented, and sleepy. Eventually, you will fall asleep, and if you have consumed enough, your nervous system will be depressed enough that your brain shuts off and you will never wake up. The main neurotransmitter system, called the γ-aminobutyric (GABA) system, regulates how active nerve transmission is. GABA's main job is to keep things calm and in control. It does this through the GABA channel, which opens selectively. Barbiturates kick this inhibitor system into overdrive and force the GABA channels to stay open for too long. When the channels are open for an extended period of time, the voltage in the brain cells is changed in a way that makes them resistant to nerve impulses. This is great if only a little bit is used, but a bit too much and your brain is no longer able to send signals. Barbiturates can act anywhere from several hours to seconds. The time frame fits—short enough to ensure a quick death with enough time to deliver some killer final lines. Olenna most likely died peacefully in a dreamless sleep after making sure Jaime would never sleep again.

CRUSHING THE SKULL

One of the most soul-crushing (and skull-crushing) deaths of season 4 was the death of Oberyn Martell. For several brief moments we all thought that he would win, and that Tyrion would escape the trial by combat with his life. That joy was all too short-lived when hubris got the better of him and he gave the Mountain the chance to crush his skull. The Mountain is undoubtedly huge—8 feet tall and about 420 pounds, according to GRRM—but would he be large and strong enough to crush a human skull? It appears after several viewings of the scene that the Mountain first jams his thumbs in Oberyn's eyes. That alone is not enough to force a brain to pop out of the skull. It appears, to me at least, that he then squeezes Oberyn's skull between his hands until it crushes and that causes his death. Even if he's not squeezing but just pushing on the bone of the eye socket the physics will be roughly the same. The physics of this isn't all that complicated, we just need to know the force needed to crush a skull and the force that the Mountain's hands could apply. This is so easy that many a science journalist has tackled it. There are actually several articles on the subject—some are good; some not so much. For this section, I went back to the original research to draw my own conclusions.

Many studies have been conducted to determine the force needed to crush a human skull, most of them by car safety groups and helmet manufacturers. There is one particular paper that is quoted in most science publications' articles about this scene. I would not recommend reading the article *ScienceAlert* published on the topic, because the units used are more than a bit of a mess. The units used in the original paper also gave me some pause, and for about the millionth time in my life I wanted to yell at a doctor, "Wrong units, minus 10 points!" (I used to teach pre-meds.) Once I figured out what the units actually were, I learned that it takes about 520 pounds of force to cause catastrophic failure in a human skull. The study only tested the skulls of 10-year-olds, but the bones of the skull have fully formed by that age, generally speaking, and although a 10-year-old's head is smaller than an adult's, they would take similar amounts of force to crush.[13] The force needed specifically to crush a skull is different from the force needed to crush other types of bones, such as the femur or the ulna. That's because the dome shape of the head adds extra strength. If you don't believe me, here's a fun experiment: Go get an egg and hold it over the sink. Try and crush it from the flat sides and see how much force it takes. Now try and crush it by pushing on the top and bottom. It takes quite a bit more force. This is due to the dome shape on the ends. A skull is no different—it takes much more force to crush it than it would to crush a flat bone like, say, a shoulder blade. Knowing that the magic crushing number is 520 pounds of force, the next step is to figure out how much force the Mountain could apply.

The Mountain only weighs 420 pounds. Even if the Mountain leaned his entire body on Oberyn's head it wouldn't crush. He'd have a hell of a headache, but no fractures. You might be saying, but wait, boxers fracture skulls all the time, why couldn't a huge man do the same? Boxers have the advantage of being in motion when their fists hit a head. That motion adds to the force applied and they can end up delivering over 1,000 pounds of force with a moving punch. The Mountain did not have this advantage. If, instead of pushing on Oberyn's skull, he had just landed one good punch, you wouldn't even be reading this section. But is anyone strong enough to crush a skull with their bare hands? According to NASA, the average man can exert up to 300 pounds of force.[14] That is still well below what would be needed to crush a skull. The Mountain would have to be about 73% stronger than the average man to be able to crush his skull. I like being able to wrap these sections up in a neat little bow saying "yes" or "no" to the question asked; unfortunately, this one is a solid "maybe." I don't know how much stronger the Mountain is than the average person. Twice as strong? One-and-a-half times as strong? He is a unique human and I assume NASA averages don't apply. I'll leave it at this: he very well might have been able to crush Oberyn's skull, but he should have just given him a solid roundhouse punch to really do the job.

Burning at the Stake

Burning at the stake is one of the oldest methods of execution, and it has also been used as a method of human sacrifice. Mance Rayder and Shireen Baratheon can collectively tell you about both. It has quite the history, and I would encourage you to look up the social and religious implications of this execution method; however, I'll be sticking to the biology of death by burning at the stake. Before you continue, I should say that I have found this to be the least enjoyable section to write, and given the chapter title and the other topics covered here, that is really saying something. In fact, I found this method of dying to be so horrible that I almost didn't write this section. But, seeing as there were two notable deaths by burning at the hands of the Red Woman, it needed to be included. When being burned you are being killed in two different ways: First, your body is becoming the fuel of the fire and is slowly consumed. Second, the fumes produced by the fire and the heat are entering your lungs and suffocating you. If you are lucky, the fumes will get you quickly. Oftentimes, the victims were not so lucky.

Whether you burn or suffocate will depend on the position of both the fire and the stake. As I mentioned in chapters 9 and 10, fire needs oxygen to burn and produces carbon monoxide, carbon dioxide, and water. When people are trapped in a burning building, they usually die from suffocation. There is little oxygen as the fire consumes it and the gas produced isn't exactly what our lungs need. When being burned at the stake, your best hope of dying by asphyxiation is for the stake and the fuel to be set up in such a way that the gases produced are trapped. This is much more likely to happen if the fuel is piled around the stake instead of under it because the wood used as fuel will trap the gases produced and keep out oxygen. Surrounded by fire, the condemned will breathe in hot carbon monoxide and carbon dioxide. The heat will burn the throat and lungs and irritate the throat, probably causing swelling. As carbon monoxide enters the bloodstream through the lungs, it pushes what oxygen there is out of the way. Hemoglobin is the protein in blood that shuttles oxygen around to the cells that need it. Unfortunately, hemoglobin likes carbon monoxide about 200 times more than oxygen. Your cells don't like CO as much as hemoglobin does, however, and without oxygen they begin to die. This quickly leads to dizziness and unconsciousness . . . if you're lucky. Even if you were able to inhale some oxygen, the CO would push it out of the way and you'd still asphyxiate. In an enclosed space, the exhaust from older cars can pump out a lethal amount of CO in about 10 minutes.[15] Shireen and Mance were just not that lucky. Besides, this wouldn't make for a good and dramatic TV shot.

When the prisoner is placed above rather than inside of the fuel and flame, death is caused by the fire, not the smoke. The smoke isn't trapped near the victim, who also has ready access to oxygen. The key to a quick death is falling unconscious as quickly as possible. When flames start at your feet that isn't easy. If you have ever burned yourself, you know it can be very painful. First, the flames break down the proteins in your skin cells, causing them to burst. Nerves are also damaged by the flames, which causes the characteristic burning sensation. Eventually, the burns will progress from first-degree skin burns to second-degree burns that kill the nerves to the point they can no longer send pain signals to the brain. From there, the burns increase in severity to third-degree burns, consuming muscles and then bones. Depending on the rate at which the fire is burning, by the time your feet are finally deadened to pain, your thighs will be experiencing first and second-degree burns. Hopefully, you will enter shock at this point. As your body is burning, your heart and breathing rates will decrease. Body structures that contain necessary fluids will break down in the flames, and you will begin to lose blood and other fluids, eventually dying of blood loss or multiple organ failure. At this point, the flames probably still haven't reached your head and granted you unconsciousness. Humans can burn for a very long time. We have a fair amount of fat, which, as discussed in chapter 9, is a pretty good fuel. Clothing can act just like a wick, turning the condemned into a human candle—and candles don't burn quickly. "It will all be over soon, Princess" couldn't be further from the truth. There are records of people burning for 45 minutes before succumbing to death. Don't die in a fire.[16]

DROWNING

"What Is Dead May Never Die" are the words of House Greyjoy, but they also appear to signify a religious ritual of this house. And people thought confirmation was a lot of work! In both the show and the book, followers of the Drowned God are purposely drowned and resuscitated to become part of the brotherhood of drowned men. It is said this is usually done soon after birth but has morphed over the years to a more normal baptism of simply dunking the child in water. But in *A Feast for Crows,* Aeron Greyjoy, aka Damphair, brings back the drowning and resuscitation. This is one of those things we've all seen happen many times on TV: someone drowns and is brought back, with the requisite coughing and throwing up water. But does that happen enough that it's reasonable to expect a priest to be able to bring back your newborn son or yourself after an intentional drowning? Turns out that, yes, it is pretty common to be able to resuscitate someone after they've drowned.

The definition of "drowning" is a bit tricky. I went into this thinking drowning meant to die by inhaling water. That's not totally accurate. Drowning is defined by the World Health Organization as experiencing respiratory impairment from submersion or immersion in liquid. When someone can no longer keep their head above water, whether that is from exhaustion or because a priest is holding their head underwater, the first response is to hold their breath as long as possible, usually about a minute. Once that's no longer possible, the person will breathe in water, try to cough it out, and breathe in more. Sometimes there's a muscle spasm in the throat that stops the inhalation of more water, but that stops as soon as the person passes out. Water in the lungs doesn't just mean that oxygen can't get in, it messes with the membranes of the alveoli, the tiny sacs in the lungs that allow oxygen to be transferred to the blood. The water gets in the way of this transfer and causes the membranes to be more permeable, which in turn allows not just oxygen but fluid to transfer into the lungs. If the person is in fresh water, osmosis causes water to be pulled into the blood. If they are in salt water, then water is pulled out of the blood and into the lungs. This is why even a little bit of water in the lungs can have a huge effect. Heart attack due to lack of oxygen to the brain quickly follows. If the person is rescued before their heart stops, it is called a nonfatal drowning; if they are not so lucky, it is a fatal drowning. If you want to survive drowning, your best bet is to drown in very cold water. Death from drowning occurs when there isn't enough oxygen to power the brain and heart. Although humans are warm-blooded, cold still has a large effect on the metabolism. If a person is drowning in cold water, their metabolism slows down and less oxygen is needed by the brain. This means that it takes much longer to die. There is a recorded case of a child being submerged in icy water for 66 minutes and recovering with no ill effects. Obviously, this is an outlier, but it's a good example of what cold can do. In the case of the Iron Islands, the water is described as being very cold. This is probably good news for those followers of the Drowned God.[17]

If a person is pulled out of the water soon enough, it's possible for the water to be cleared from their lungs, at which point the alveoli membranes to go back to doing their job and the brain gets its much-needed oxygen. If a person is rescued after being submerged for 5 minutes or less, there's a 90% chance they will be just fine. Submersion for 6 minutes still has a 44% chance of survival with minimal brain impairment. The two goals in reviving a drowning victim are to get the heart going (or keep it going) and get the lungs to the point of transferring oxygen to the blood again. CPR does both. Those trained in drowning rescue are taught to try rescue breathing while the victim is still in the water instead of waiting to get them to land. In the water, it's pretty tough to do chest compressions, and really, they aren't recommended anyway unless it's clear the heart has stopped. Chest compressions make the victim vomit 86% of the time, and the Heimlich maneuver does pretty much the same. Mouth-to-mouth resuscitation pushes oxygen into the lungs and makes sure that the alveoli not compromised by water are getting enough oxygen. Once things are back up and running, any water remaining in the lungs is coughed up and the victim eventually goes back to normal.

In the case of the Iron Islanders, this is all really good news. The water is described as very cold, which certainly helps the survival rate. In addition, they probably didn't have a way to make sure they actually killed their drowning victims. Hopefully, this means that the one being "baptized" is removed from the water before their heart stops. In this case, with a little mouth-to-mouth, the initiate will revive and return to normal to serve the Drowned God. I can't say this would be the religion or initiation method I would choose, but the science seems to indicate there's about a 90% chance of survival, which is better than I thought it would be the first time I read about Damphair trying this out.

SO, WHAT TYPE OF JUSTICE WOULD YOU PICK?

As I said at the beginning of this chapter, it is difficult to research and write something like this without imagining myself as an unfortunate character in *Game of Thrones*. To be clear, the odds of surviving the series are vanishingly slim for those who want to play the game and survive. Jon seems to run toward death as if its name is Ygritte, and yet he keeps coming back. It's uncanny. Personally, I had to take a break in the middle of each of the books, as I was having nightmares about being killed in the increasingly brutal ways devised by GRRM. As he has made clear time and again, no one is safe. The natural question, then, is: If I got drafted into Westeros, which way would I pick to go? This was up for much debate both in the office and at various social gatherings. For months I was a lot of fun at parties. The conclusion I've come to personally is that I'd pick pretty much anything but burning. Poisoning by anything but barbiturates is second on my "would not recommend" list. Most others lead to quick unconsciousness and death. I hope, however, that I could go out with a great gut punch like Lady Olenna, or, as Tyrion once hoped for himself, simply die in my own bed with a belly full of wine, and, well . . . you know the rest.

Epilogue

I hope you enjoyed reading about the science behind the cultural phenomenon that is *Game of Thrones*. The world created by George R. R. Martin, D. B. Weiss, and David Benioff is deep and rich, and I've enjoyed highlighting some of the science behind it. More than anything, I hope this is a starting point for you to learn more about whatever science you found the most interesting. I hope this book leads to many a Wikipedia rabbit hole and a more in-depth understanding of how our world works.

I have found that stories—and fantasy stories in particular—are best when they ask you to believe in the fantastical but don't require you to jump through too many mental hoops. You want to relate to the world and its characters. If you are asked to ignore too many laws of physics or to believe in too much questionable biology, that relatability is lost. The rules of reality can be stretched in a fantasy world, but they need to be internally consistent. If they aren't, even the most engrossed viewer will be forced to pause and say, "Huh?"

Before writing this book, I knew I could get lost in the world of Westeros, but I wasn't fully aware of the role science was playing in that. George R. R. Martin was able to create a world that was internally consistent and just a hair beyond reality. Why couldn't we live on a planet where seasons are hard to track? After looking at the science of it, it's a wonder we don't! If bones were just a bit stronger, we could have flying dragons. Damascus steel might not kill a White Walker or be forged with dragon fire, but it certainly earned its legendary reputation. And this is why the world of George R. R. Martin is so compelling: it's so close to ours yet just out of reach.

After living in the real world for so many years, you have an inherent, if not totally mathematical, understanding of the laws of the universe. You know how things are supposed to fly or fall, how a ship's shape affects its speed, and why you'd rather be beheaded by guillotine than an axe. The best fictional worlds leave the things you understand on a gut instinct level intact and build on them. But, if a story starts playing with what we understand to be true in our gut, it throws us out of the story. A great example of this can be seen in comparing a scene from *Speed* and a scene from *Frozen*. In *Frozen*, the reindeer, Sven, successfully jumps across a large crevasse to carry Anna to safety. It's a big jump for a reindeer but not for the audience. Most viewers accepted that Sven could make the jump. By contrast, in the movie *Speed*, there is the famous scene of the bus jumping across the gap in the bridge. Numerous physics students have been given the task of figuring out if the bus could make it. The difference between *Speed* and *Frozen* is that on a gut level we know Anna and Sven will be fine, but something seems wrong with the bus. If you do the math, Sven would have made it and the bus would not. Our gut knows this even without vectors and algebra. The incorrect science of the bus leap pulls us out of the story even if only a little bit. For the most part, GRRM did an excellent job of keeping most laws of science relatively intact, which helped keep readers and viewers engrossed. The characters and story remained the focus because we as viewers and readers weren't constantly saying, "Huh?"

Because of how well the show incorporates science, analyzing the scientific aspects of *Game of Thrones* is a good jumping-off point for broader discussions of scientific topics. I've spent my life working as a science communicator, and the number one rule in communication is to tell a story. Tell people why they should care about the information you are presenting. The most compelling arguments for science are the ones where you know the players and care about what happens. By using a story such as *Game of Thrones* to start a discussion about science, your audience is already engaged.

There are several different ways to use stories to convey information about science. First, scientists telling personal stories about their journeys and why they chose their field of research can draw an audience in a different way than simply presenting their findings does. This is one of the most effective ways for research to be shared. People want to connect research with real life. Another way of incorporating storytelling into science is to write science-based stories. This is what I do with the *Spectra* comic book series. Following the adventures of a science superhero teaches the reader some science along the way. A third method of presenting science through stories is what I've done in this book, using an existing story as a backdrop for teaching. This gives the science concepts some connection to a world, even if it's not the real world. It makes both the science and the fictional world seem more real, more tangible.

The idea of using pop culture such as *Game of Thrones* to teach science isn't new. You can find several Hollywood science books on the market, and many sci-fi shows have an associated "science of" book. Though the concept isn't new, it is effective. There are physics courses based entirely on movies or comic books. The average science curriculum doesn't have a narrative thread woven throughout the course; it simply moves from one topic to another. Anchoring the course in a story or world keeps students connected and engaged. I hope that you found that connection in this book.

If you are a scientist reading this book, I hope you walk away with new and interesting ways to explain the work you do. I hope you will use stories in your teaching, talks, and papers. If you are a *Game of Thrones* enthusiast who also likes science, I hope this book has made the world of Westeros feel more real and more alive. I also hope you will be more likely to look at the real world and other fictional worlds in a new way. I hope everyone who enjoyed this book will now go off to learn more about some of the topics presented here and spend many hours on Twitter and Reddit discussing the details of others' imaginations.

Notes

Chapter 1: Winter Is Coming—Or Is It?

1. I'll let the site's creator explain: "I created this site as a joke to poke fun at the fact that, every time the weather gets cold (especially in certain parts of the country), stores seem to quickly run out of milk, bread, and eggs . . . which happen to be the ingredients for making French toast!" See Shawn C. Reed, "The National French Toast Index," http://frenchtoastalert.com/.

2. Mary Lou Whitehorne, "Why Earth Is Closest to Sun in Dead of Winter," *Space.com*, January 2, 2007. https://www.spacc.com/3304-earth-closest-sun-dead-winter.html.

3. Tim Sharp, "How Far Is Earth from the Sun?" *Space.com*, October 18, 2017, https://www.space.com/17081-how-far-is-earth-from-the-sun.html.

4. "Washington DC, USA: Annual Weather Averages," *HolidayWeather.com*, http://www.holiday-weather.com/washington_dc/averages/.

5. NASA, "Earth at Perihelion," January 3, 2001, https://science.nasa.gov/science-news/science-at-nasa/2001/ast04jan_1.

6. J. Laskar, F. Joutel, and P. Robutel, "Stabilization of the Earth's Obliquity by the Moon," *Nature* 361, no. 6413 (1993): 615–617, https://doi.org/10.1038/361615a0; J. Laskar and P. Robutel, "The Chaotic Obliquity of the Planets," *Nature* 361, no. 6413 (1993): 608–612, https://doi.org/10.1038/361608a0.

7. J. D. Hays, John Imbrie, and N. J. Shackleton, "Variations in the Earth's Orbit: Pacemaker of the Ice Ages," *Science* 194, no. 4270 (1976): 1121–1132. https://doi.org/10.1126/science.194.4270.1121.

8. Figure 1 in Veselin Kostov, Daniel Allan, Nikolaus Hartman, Scott Guzewich, and Justin Rogers, "Winter Is Coming," *arXiv*, April 1, 2013, https://arxiv.org/abs/1304.0445.

CHAPTER 2: AND NOW MY WATCH BEGINS

1. Lasse Makkonen and Maria Tikanmäki, "Modeling the Friction of Ice," *Cold Regions Science and Technology* 102 (2014): 84–93, https://doi.org/10.1016/j.cold regions.2014.03.002.

2. W. D. Kingery, "Ice Alloys: For Arctic Operations Ice and Snow Can Be Improved as Structural Materials by Appropriate Alloying," *Science* 134, no. 3473 (1961): 164–168, https://doi.org/10.1126/science.134.3473.164.

3. M. F. Perutz, "A Description of the Iceberg Aircraft Carrier and the Bearing of the Mechanical Properties of Frozen Wood Pulp upon Some Problems of Glacier Flow," *Journal of Glaciology* 1, no. 3 (1948): 95–104, https://doi.org/10.1017/s002214 3000007796.

4. D. Tabor and J. C. F. Walker, "Creep and Friction of Ice," *Nature* 228, no. 5267 (1970): 137–139, https://doi.org/10.1038/228137a0.

CHAPTER 3: NORTH OF THE WALL

1. James D. Hardy, Eugene F. Du Bois, and G. F. Soderstrom, "Basal Metabolism, Radiation, Convection and Vaporization at Temperatures of 22 to 35°C," *The Journal of Nutrition* 15, no. 5 (1938): 477–497, https://doi.org/10.1093/jn/15.5.477.

2. Brian Phillips, "Out in the Great Alone," *ESPN*, May 5, 2013, http://www.espn .com/espn/grantland/story/_/id/9175394/out-great-alone.

3. Kenneth R. Holmes, "Thermal Properties," http://users.ece.utexas.edu/~valvano/ research/Thermal.pdf.

4. L. Ye, J. Wu, P. Cohen, L. Kazak, M. J. Khandekar, M. P. Jedrychowski, X. Zeng, S. P. Gygi, and B. M. Spiegelman, "Fat Cells Directly Sense Temperature to Activate Thermogenesis," *Proceedings of the National Academy of Sciences* 110, no. 30 (2013): 12480–12485, https://doi.org/10.1073/pnas.1310261110.

5. Heather E. M. Liwanag, Annalisa Berta, Daniel P. Costa, Masako Abney, and Terrie M. Williams, "Morphological and Thermal Properties of Mammalian Insulation: The Evolution of Fur for Aquatic Living," *Biological Journal of the Linnean Society* 106, no. 4 (2012): 926–939, https://doi.org/10.1111/j.1095-8312.2012.01900.x.

6. Priscilla Simonis, Mourad Rattal, El Mostafa Oualim, Azeddine Mouhse, and Jean-Pol Vigneron, "Radiative Contribution to Thermal Conductance in Animal Furs and Other Woolly Insulators," *Optics Express* 22, no. 2 (2014): 1940, https://doi .org/10.1364/oe.22.001940.

7. Science History Institute, "Robert W. Gore," January 16, 2018, https://www.science history.org/historical-profile/robert-w-gore.

8. The North Face, "Insulation Opens Up: TNF 'Ventrix' Jacket Dissected," *GearJunkie*, October 29, 2017, https://gearjunkie.com/the-north-face-ventrix.

9. Engineering ToolBox, "Thermal Conductivity of Common Materials and Gases," 2003, https://www.engineeringtoolbox.com/thermal-conductivity-d_429.html.

10. Thea Pretorius, Gerald K. Bristow, Alan M. Steinman, and Gordon G. Giesbrecht, "Thermal Effects of Whole Head Submersion in Cold Water on Nonshivering Humans," *Journal of Applied Physiology* 101, no. 2 (2006): 669–675, https://doi.org/10.1152/japplphysiol.01241.2005.

11. Gord Pyzer, "Ice-fishing Warm-up: Want to Know How Fast Your Lake Will Freeze?" *Outdoor Canada*, December 14, 2016, http://www.outdoorcanada.ca/How_Fast_Does_Your_Lake_Make_Ice.

CHAPTER 4: WHITE WALKERS, ZOMBIES, PARASITES, AND STATISTICS

1. http://zombieresearchsociety.com

2. Charles E. Rupprecht, "Rhabdoviruses: Rabies Virus," in *Medical Microbiology*, 4th ed., ed. Samuel Baron (Galveston, TX: University of Texas Medical Branch at Galveston, 1996), https://www.ncbi.nlm.nih.gov/books/NBK8618/.

3. A. Tipold, M. Vandevelde, and A. Jaggy, "Neurological Manifestations of Canine Distemper Virus Infection." *Journal of Small Animal Practice* 33, no. 10 (10 1992): 466–470, https://doi.org/10.1111/j.1748-5827.1992.tb01024.x.

4. Ram Gal and Frederic Libersat, "A Wasp Manipulates Neuronal Activity in the Sub-Esophageal Ganglion to Decrease the Drive for Walking in Its Cockroach Prey," *PLoS ONE* 5, no. 4 (2010), https://doi.org/10.1371/journal.pone.0010019.

5. Maridel A. Fredericksen, Yizhe Zhang, Missy L. Hazen, Raquel G. Loreto, Colleen A. Mangold, Danny Z. Chen, and David P. Hughes, "Three-dimensional Visualization and a Deep-learning Model Reveal Complex Fungal Parasite Networks in Behaviorally Manipulated Ants," *Proceedings of the National Academy of Sciences* 114, no. 47 (2017): 12590–12595, https://doi.org/10.1073/pnas.1711673114.

6. A. Vyas, S.-K. Kim, N. Giacomini, J. C. Boothroyd, and R. M. Sapolsky, "Behavioral Changes Induced by Toxoplasma Infection of Rodents Are Highly Specific to Aversion of Cat Odors," *Proceedings of the National Academy of Sciences* 104, no. 15 (2007): 6442–6447, https://doi.org/10.1073/pnas.0608310104.

7. J. P. Webster, P. H. L. Lamberton, C. A. Donnelly, and E. F. Torrey, "Parasites as Causative Agents of Human Affective Disorders? The Impact of Anti-psychotic, Mood-stabilizer and Anti-parasite Medication on *Toxoplasma gondii*'s Ability to Alter Host Behaviour," *Proceedings of the Royal Society B: Biological Sciences* 273, no. 1589 (2006): 1023–1030, https://doi.org/10.1098/rspb.2005.3413.

8. Timothy Verstynen and Bradley Voytek, *Do Zombies Dream of Undead Sheep? A Neuroscientific View of the Zombie Brain* (Princeton, NJ: Princeton University Press, 2014).

9. Alexander A. Alemi, Matthew Bierbaum, Christopher R. Myers, and James P. Sethna, "You Can Run, You Can Hide: The Epidemiology and Statistical Mechanics of Zombies," *Physical Review E* 92, no. 5 (2015), https://doi.org/10.1103/physreve.92.052801.

10. Robert P. Sharp, "Suitability of Ice for Aircraft Landings," *Transactions, American Geophysical Union* 28, no. 1 (1947): 111, https://doi.org/10.1029/tr028i001p00111.

11. Gord Pyzer, "Ice-fishing Warm-up: Want to Know How Fast Your Lake Will Freeze?" *Outdoor Canada*, December 14, 2016, http://www.outdoorcanada.ca/How_Fast_Does_Your_Lake_Make_Ice.

Chapter 5: Regular Steel, Made in Pittsburgh

1. Huang Yi, Guoping Xu, Huigao Cheng, Junshi Wang, Yinfeng Wan, and Hui Chen, "An Overview of Utilization of Steel Slag," *Procedia Environmental Sciences* 16 (2012): 791–801, https://doi.org/10.1016/j.proenv.2012.10.108.

2. Nikolaas J. Van der Merwe and Donald H. Avery, "Pathways to Steel: Three Different Methods of Making Steel from Iron Were Developed by Ancient Peoples of the Mediterranean, China, and Africa," *American Scientist* 70, no. 2 (1982): 146–155, https://www.jstor.org/stable/27851346.

3. Robert Maddin, James D. Muhly, and Tamara S. Wheeler, "How the Iron Age Began," *Scientific American* 237, no. 4 (1977): 122–131, https://doi.org/10.1038/scientificamerican1077-122.

4. V. M. Chernov, B. K. Kardashev, and K. A. Moroz, "Low-temperature embrittlement and fracture of metals with different crystal lattices—Dislocation mechanisms," *Nuclear Materials and Energy* 9 (2016): 496–501, https://doi.org/10.1016/j.nme.2016.02.002.

Chapter 6: Valyrian Steel, Made in Damascus

1. A. A. Levin, D. C. Meyer, M. Reibold, W. Kochmann, N. Pätzke, and P. Paufler, "Microstructure of a Genuine Damascus Sabre," *Crystal Research and Technology* 40, no. 9 (2005): 905–916, https://doi.org/10.1002/crat.200410456.

2. Oleg D. Sherby, "Ultrahigh Carbon Steels, Damascus Steels and Ancient Blacksmith," *ISIJ International* 39, no. 7 (1999): 637–648, https://doi.org/10.2355/isijinternational.39.637.

3. J. D. Verhoeven, "Damascus Steel, Part I: Indian Wootz Steel," *Metallography* 20, no. 2 (1987): 145–151, https://doi.org/10.1016/0026-0800(87)90026-7.

4. J. D. Verhoeven, A. H. Pendray, and W. E. Dauksch, "The Continuing Study of Damascus Steel: Bars from the Alwar Armory," *Jom* 56, no. 9 (2004): 17–20, https://doi.org/10.1007/s11837-004-0193-4.

5. T. H. Maugh, "A Metallurgical Tale of Irony," *Science* 215, no. 4529 (1982): 153, https://doi.org/10.1126/science.215.4529.153.

6. J. D. Verhoeven, H. H. Baker, D. T. Peterson, H. F. Clark, and W. M. Yater, "Damascus Steel, Part III: The Wadsworth–Sherby Mechanism," *Materials Characterization* 24, no. 3 (1990): 205–127, https://doi.org/10.1016/1044-5803(90)90052-l.

7. Oleg D. Sherby, "Ultrahigh Carbon Steels, Damascus Steels and an Ancient Blacksmith," *ISIJ International* 39, no. 7 (1999): 637–648, https://doi.org/10.2355/isijinternational.39.637.

8. John D. Verhoeven, "The Mystery of Damascus Blades," *Scientific American* 284, no. 1 (2001): 74–79, https://doi.org/10.1038/scientificamerican0101-74.

9. Levin et al., "Microstructure of a Genuine Damascus Sabre."

10. M. Reibold, P. Paufler, A. A. Levin, W. Kochmann, N. Pätzke, and D. C. Meyer, "Carbon Nanotubes in an Ancient Damascus Sabre," *Nature* 444, no. 7117 (2006): 286, https://doi.org/10.1038/444286a.

11. D. Golberg, M. Mitome, Ch. Müller, C. Tang, A. Leonhardt, and Y. Bando, "Atomic Structures of Iron-based Single-crystalline Nanowires Crystallized inside Multi-walled Carbon Nanotubes as Revealed by Analytical Electron Microscopy," *Acta Materialia* 54, no. 9 (2006): 2567–2576, https://doi.org/10.1016/j.actamat.2006.01.040.

12. Lei Ni, Keiji Kuroda, Ling-Ping Zhou, Tokushi Kizuka, Keishin Ohta, Kiyoto Matsuishi, and Junji Nakamura, "Kinetic Study of Carbon Nanotube Synthesis over Mo/Co/MgO Catalysts," *Carbon* 44, no. 11 (2006): 2265–2272, https://doi.org/10.1016/j.carbon.2006.02.031.

13. Maugh, "A Metallurgical Tale of Irony."

14. T. Dumitrica, M. Hua, and B. I. Yakobson, "Symmetry-, Time-, and Temperature-dependent Strength of Carbon Nanotubes," *Proceedings of the National Academy of Sciences* 103, no. 16 (2006): 6105–6109, https://doi.org/10.1073/pnas.0600945103.

CHAPTER 7: DRAGON BIOLOGY

1. Brian Rohrig, "Chilling Out, Warming Up: How Animals Survive Temperature Extremes," *ChemMatters Online*, October 2013, https://www.acs.org/content/acs/en/education/resources/highschool/chemmatters/past-issues/archive-2013-2014/animal-survival-in-extreme-temperatures.html.

2. "Boeing Test Flight Statistics," http://www.dept.aoe.vt.edu/~mason/Mason_f/747.IN.

3. Brown University Aeromechanics & Evolutionary Morphology Lab, http://www.brown.edu/Departments/EEB/EML/.

4. S. M. Swartz, J. Iriarte-Diaz, D. K. Riskin, and K. S. Breuer, "A Bird? A Plane? No, It's a Bat: An Introduction to the Biomechanics of Bat Flight," in *Evolutionary History of Bats: Fossils, Molecules, and Morphology* (Cambridge: Cambridge University Press, 2012), 317–352.

5. Swartz et al., "A Bird? A Plane? No, It's a Bat."

6. F. T. Mujires, L. C. Johansson, R. Barfield, M. Wolf, G. R. Spedding, and A. Hedenström, "Leading-Edge Vortex Improves Lift in Slow-Flying Bats," *Science* 319, no. 5867 (2008): 1250–1253, https://doi.org/10.1126/science.1153019.

7. U. M. Lindhe Norberg and Y. Winter, "Wing Beat Kinematics of a Nectar-feeding Bat, *Glossophaga soricina*, Flying at Different Flight Speeds and Strouhal Numbers," *Journal of Experimental Biology* 209, no. 19 (2006): 3887–3897, https://doi.org/10.1242/jeb.02446.

8. M. P. Witton and M. B. Habib, "On the Size and Flight Diversity of Giant Pterosaurs, the Use of Birds as Pterosaur Analogues and Comments on Pterosaur Flightlessness," *PLoS ONE* 5, no. 11 (2010): e13982, https://doi.org/10.1371/journal.pone.0013982.

9. Witton and Habib, "The Size and Diversity of Giant Pterosaurs."

10. J. S. Findley, E. H. Studier, and D. E. Wilson, "Morphologic Properties of Bat Wings," *Journal of Mammalogy* 53, no. 3 (1972): 429–444, https://doi.org/10.2307/1379035.

11. U. M. Norberg, "Aerodynamics, Kinematics, and Energetics of Horizontal Flapping Flight in the Long-eared Bat *Plecotus auratus*," *Journal of Experimental Biology* 65, no. 1 (1976): 179–212.

CHAPTER 8: HOW TO KILL A WHITE WALKER

1. John C. Mauro and Edgar D. Zanotto, "Two Centuries of Glass Research: Historical Trends, Current Status, and Grand Challenges for the Future," *International Journal of Applied Glass Science* 5, no. 3 (2014): 313–327, https://doi.org/10.1111/ijag.12087; Zanotto, Edgar D. "Glass Crystallization Research—A 36-Year Retrospective. Part I, Fundamental Studies," *International Journal of Applied Glass Science* 4, no. 2 (2013): 105–116, https://doi.org/10.1111/ijag.12022.

2. H. J. Fecht, "Thermodynamic Properties of Amorphous Solids—Glass Formation and Glass Transition—(*Overview*)," *Materials Transactions, JIM* 36, no. 7 (1995): 777–793, https://doi.org/10.2320/matertrans1989.36.777; Michael P. Marder, *Condensed Matter Physics* (Malden, MA: Wiley, 2004).

3. Prabhat K. Gupta, "Non-crystalline Solids: Glasses and Amorphous Solids," *Journal of Non-Crystalline Solids* 195, no. 1–2 (1996): 158–164, https://doi.org/10.1016/0022-3093(95)00502-1.

4. Mark Ford, "How Is Tempered Glass Made?" *Scientific American*, January 22, 2001, https://www.scientificamerican.com/article/how-is-tempered-glass-mad/.

5. Zeeya Merali, "This 1,600-Year-Old Goblet Shows That the Romans Were Nanotechnology Pioneers," *Smithsonian Magazine*, September 2013, https://www.smithsonianmag.com/history/this-1600-year-old-goblet-shows-that-the-romans-were-nanotechnology-pioneers-787224/.

6. J. E. Ericson, A. Makishima, J. D. Mackenzie, and R. Berger, "Chemical and Physical Properties of Obsidian: A Naturally Occuring Glass," *Journal of Non-Crystalline Solids* 17, no. 1 (1975): 129–142, https://doi.org/10.1016/0022-3093(75)90120-9.

7. R. H. Kropschot and R. P. Mikesell, "Strength and Fatigue of Glass at Very Low Temperatures," *Journal of Applied Physics* 28, no. 5 (1957): 610–614, https://doi.org/10.1063/1.1722812.

CHAPTER 9: HARRENHAL

1. Alan Alda, "The Flame Challenge," *Science* 335, no. 6072: 1019, https://doi.org/10.1126/science.1220619

2. International Association of Fire Chiefs, International Association of Arson Investigators, and National Fire Protection Association, *Fire Investigator: Principles and Practice to NFPA 921 and 1033*, 5th ed. (Burlington, MA: Jones & Bartlett Learning, 2018).

3. C. Grainger, T. Clarke, S. M. Mcginn, M. J. Auldist, K. A. Beauchemin, M. C. Hannah, G. C. Waghorn, H. Clark, and R. J. Eckard, "Methane Emissions from Dairy Cows Measured Using the Sulfur Hexafluoride (SF6) Tracer and Chamber Techniques," *Journal of Dairy Science* 90, no. 6 (2007): 2755–2766, https://doi.org/10.3168/jds.2006-697.

4. U.S. Energy Information Administration, "Chinese Coal-fired Electricity Generation Expected to Flatten as Mix Shifts to Renewables," *Today in Energy* (blog), September 27, 2017, https://www.eia.gov/todayinenergy/detail.php?id=33092.

5. Philip E. Mason, Frank Uhlig, Václav Vaněk, Tillmann Buttersack, Sigurd Bauerecker, and Pavel Jungwirth, "Coulomb Explosion during the Early Stages of the Reaction of Alkali Metals with Water," *Nature Chemistry* 7, no. 3 (2015): 250–254, https://doi.org/10.1038/nchem.2161.

6. Majid Minary-Jolandan and Min-Feng Yu, "Nanoscale Characterization of Isolated Individual Type I Collagen Fibrils: Polarization and Piezoelectricity," *Nanotechnology* 20, no. 8 (2009): 085706, https://doi.org/10.1088/0957-4484/20/8/085706.

7. Becky Oskin, "Deep-Sea Worms Can't Take the Heat," *Live Science*, May 29, 2013, https://www.livescience.com/34835-heat-limit-for-life.html.

8. Neil DeGrasse Tyson (@neiltyson), "Intriguing thermal physics in #GameOfThrones: BlueDragon breath would be at least a factor of 3X hotter than RedDragon breath," Twitter, September 24, 2017, 5:46 p.m., https://twitter.com/neiltyson/status/912070876950122496.

9. Mostafa Shazly, Vikas Prakash, and Bradley A. Lerch, "High Strain-rate Behavior of Ice under Uniaxial Compression," *International Journal of Solids and Structures* 46, no. 6 (2009): 1499–1515, https://doi.org/10.1016/j.ijsolstr.2008.11.020.

Chapter 10: The Battle of the Blackwater

1. Jennifer Latson, "The Burning River That Started a Revolution," *Time*, June 22, 2015, http://time.com/3921976/cuyahoga-fire/.

2. Robert M. Neer, *Napalm: An American Biography* (Cambridge, MA: Harvard University Press, 2015).

3. Simon Cotton, "Napalm," November 13, 2013, in *Chemistry in Its Element*, produced by the Royal Society of Chemistry, podcast, MP3 audio, 5:39, https://www.chemistryworld.com/podcast/napalm/6784.article; Darion Fontaine, "The Chemistry of Napalm," *Chemistry Is Life*, http://www.chemistryislife.com/the-chemistry-of-napalm.

4. John Pike, "M15 White Phosphorus Grenade," *FAS Military Analysis Network*, September 12, 1998, https://fas.org/man/dod-101/sys/land/m15.htm; Neer, *Napalm*.

5. National Institute for Occupational Safety and Health (NIOSH), "Trimethyl Borate," Centers for Disease Control and Prevention, July 1, 2014, https://web.archive.org/web/20170614234300/https://www.cdc.gov/niosh/ipcsneng/neng0593.html.

6. J. R. Partington, *A History of Greek Fire and Gunpowder* (Cambridge: W. Heffer & Sons, 1960).

7. Alex Roland, "Secrecy, Technology, and War: Greek Fire and the Defense of Byzantium, 678–1204," *Technology and Culture* 33, no. 4 (1992): 655, https://doi.org/10.2307/3106585.

Chapter 11: Houses Targaryen and Lannister

1. Nic Koltzoff, "The Structure of the Chromosomes in the Salivary Glands of Drosophila," *Science* 80, no. 2075 (1934): 312–313, https://doi.org/10.1126/science.80.2075.312.

2. David Botstein and Neil Risch, "Discovering Genotypes Underlying Human Phenotypes: Past Successes for Mendelian Disease, Future Approaches for Complex Disease," *Nature Genetics* 33 (2003): 228–237, https://doi.org/10.1038/ng1090.

3. Sewall Wright, *Collected Scientific Papers 1929–1946 / Sewall Wright*, 1929.

4. A. Helgason, S. Palsson, D. F. Guthbjartsson, T. Kristjansson, and K. Stefansson, "An Association between the Kinship and Fertility of Human Couples," *Science* 319, no. 5864 (2008): 813–816, https://doi.org/10.1126/science.1150232.

5. R. Chris Fraley and Michael J. Marks, "Westermarck, Freud, and the Incest Taboo: Does Familial Resemblance Activate Sexual Attraction?" *Personality and Social Psychology Bulletin* 36, no. 9 (2010): 1202–1212, https://doi.org/10.1177/0146167210377180.

6. Arthur P. Wolf, *Sexual Attraction and Childhood Association: A Chinese Brief for Edward Westermarck* (Stanford: Stanford University Press, 1995).

7. R. McDermott, D. Tingley, J. Cowden, G. Frazzetto, and D. D. P. Johnson, "Monoamine Oxidase A Gene (MAOA) Predicts Behavioral Aggression following Provocation," *Proceedings of the National Academy of Sciences* 106, no. 7 (2009): 2118–2123, https://doi.org/10.1073/pnas.0808376106.

CHAPTER 12: WE DO NOT SOW

1. R. G. Grant, *Battle at Sea: 3,000 Years of Naval Warfare* (London: Dorling Kindersley, 2008).

2. W. H. Bell, "The Influence of Turbulence on Drag," *Ocean Engineering* 6, no. 3 (1979): 329–340, https://doi.org/10.1016/0029-8018(79)90021-0.

3. Bryon D. Anderson, "The Physics of Sailing," *Physics Today* 61, no. 2 (2008): 38–43. https://doi.org/10.1063/1.2883908.

4. Tamela Maciel, "The Physics of Sailing: How Does a Sailboat Move Upwind?" *Physics Buzz* (blog), May 12, 2015, http://physicsbuzz.physicscentral.com/2015/05/the-physics-of-sailing-how-does.html.

5. Ian Randall, "Research & Results," *RANDALLfoil*, https://hydrofoiloar.blogspot.com/p/documents.html.

6. Dava Sobel, *Longitude: The True Story of a Lone Genius Who Solved the Greatest Scientific Problem of His Time* (New York: Walker & Company, 2007).

7. B. Zumreoglukaran, "The Coordination Chemistry of Vitamin C: An Overview," *Coordination Chemistry Reviews* 250, no. 17–18 (2006): 2295–2307, https://doi.org/10.1016/j.ccr.2006.03.002.

8. Lynne Goebel, "Scurvy," *Medscape*, October 24, 2017, https://emedicine.medscape.com/article/125350-overview.

CHAPTER 13: THE KING'S JUSTICE

1. Marcus E. Raichle, "The Pathophysiology of Brain Ischemia," *Annals of Neurology* 13, no. 1 (1983): 2–10, https://doi.org/10.1002/ana.410130103.

2. Clementina M. van Rijn, Hans Krijnen, Saskia Menting-Hermeling, and Anton M. L. Coenen, "Decapitation in Rats: Latency to Unconsciousness and the 'Wave of Death,'" *PLoS ONE* 6, no. 1 (2011), https://doi.org/10.1371/journal.pone.0016514.

3. F. R. W. van de Goot, R. L. ten Berge, and R. Vos, "Molten Gold Was Poured Down His Throat until His Bowels Burst," *Journal of Clinical Pathology* 56, no. 2 (2003): 157, https://doi.org/10.1136/jcp.56.2.157.

4. Haiyan Li, Jesse Ruan, Zhonghua Xi, Hao Wang, and Wengling Liu, "Investigation of the Critical Geometric Characteristics of Living Human Skulls Utilising Medical Image Analysis Techniques," *International Journal of Vehicle Safety* 2, no. 4 (2007): 345–367, https://doi.org/10.1504/IJVS.2007.016747.

5. Sabrina B. Sholts, Sebastian K. T. S. Wärmländer, Louise M. Flores, Kevin W. P. Miller, and Phillip L. Walker, "Variation in the Measurement of Cranial Volume and Surface Area Using 3D Laser Scanning Technology," *Journal of Forensic Sciences* 55, no. 4 (2010): 871–876, https://doi.org/10.1111/j.1556-4029.2010.01380.x.

6. James L. Luke, Donald T. Reay, John W. Eisele, and Harry J. Bonnell, "Correlation of Circumstances with Pathological Findings in Asphyxial Deaths by Hanging: A Prospective Study of 61 Cases from Seattle, WA," *Journal of Forensic Sciences* 30, no. 4 (1985), https://doi.org/10.1520/jfs11055j.

7. Mahmoud Rayes, Monika Mittal, Setti S. Rengachary, and Sandeep Mittal, "Hangman's Fracture: A Historical and Biomechanical Perspective," *Journal of Neurosurgery: Spine* 14, no. 2 (2011): 198–208, https://doi.org/10.3171/2010.10.spine09805.

8. Department of the Army, *Procedure for Military Executions* (Washington, DC: U.S. War Office, 1947), https://www.loc.gov/rr/frd/Military_Law/pdf/procedure_dec-1947.pdf.

9. Albert Pierrepoint, *Executioner: Pierrepoint* (London: Hodder and Stoughton, 1977).

10. Rayes et al., "Hangman's Fracture."

11. Deborah Blum, *The Poisoner's Handbook: Murder and the Birth of Forensic Medicine in Jazz Age New York* (London: Penguin Books, 2011).

12. Daniel M. Lajoie, Pamela A. Zobel-Thropp, Vlad K. Kumirov, Vahe Bandarian, Greta J. Binford, and Matthew H. J. Cordes, "Phospholipase D Toxins of Brown Spider Venom Convert Lysophosphatidylcholine and Sphingomyelin to Cyclic Phosphates," *PLoS ONE* 8, no. 8 (2013), https://doi.org/10.1371/journal.pone.0072372.

13. Tobias A. Mattei, Brandon J. Bond, Carlos R. Goulart, Chris A. Sloffer, Martin J. Morris, and Julian J. Lin, "Performance Analysis of the Protective Effects of Bicycle Helmets during Impact and Crush Tests in Pediatric Skull Models," *Journal of Neurosurgery: Pediatrics* 10, no. 6 (2012): 490–497, https://doi.org/10.3171/2012.8.peds12116.

14. NASA, "Man-Systems Integration Standards, Volume I: Strength," sections 4.9.3–4.9.5, July 1995 https://msis.jsc.nasa.gov/sections/section04.htm#Figure 4.9.3–5.

15. T. Meredith and A. Vale, "Carbon Monoxide Poisoning," *BMJ* 296, no. 6615 (1988): 77–79, https://doi.org/10.1136/bmj.296.6615.77.

16. Lars H. Evers, Dhaval Bhavsar, and Peter Mailänder, "The Biology of Burn Injury," *Experimental Dermatology* 19, no. 9 (2010): 777–783, https://doi.org/10.1111/j.1600-0625.2010.01105.x.

17. David Szpilman, Joost J. L. M. Bierens, Anthony J. Handley, and James P. Orlowski, "Drowning," *New England Journal of Medicine* 366, no. 22 (2012): 2102–2110, https://doi.org/10.1056/nejmra1013317.

Index